REVOLUTIONIZING PEDAGOGY

Marxism and Education

This series assumes the ongoing relevance of Marx's contributions to critical social analysis and aims to encourage continuation of the development of the legacy of Marxist traditions in and for education. The remit for the substantive focus of scholarship and analysis appearing in the series extends from the global to the local in relation to dynamics of capitalism and encompasses historical and contemporary developments in political economy of education as well as forms of critique and resistances to capitalist social relations. The series announces a new beginning and proceeds in a spirit of openness and dialogue within and between Marxism and education, and between Marxism and its various critics. The essential feature of the work of the series is that Marxism and Marxist frameworks are to be taken seriously, not as formulaic knowledge and unassailable methodology but critically as inspirational resources for renewal of research and understanding, and as support for action in and upon structures and processes of education and their relations to society. The series is dedicated to the realization of positive human potentialities as education and, thus, with Marx, to our education as educators.

Renewing Dialogues in Marxism and Education: Openings
Edited by Anthony Green, Glenn Rikowski, and Helen Raduntz

Critical Race Theory and Education: A Marxist Response
Mike Cole

Revolutionizing Pedagogy: Education for Social Justice within and beyond Global Neoliberalism
Edited by Sheila Macrine, Peter McLaren, and Dave Hill

Revolutionizing Pedagogy

Education for Social Justice Within and Beyond Global Neo-Liberalism

Edited by
Sheila Macrine, Peter McLaren, and
Dave Hill

palgrave
macmillan

First published in 2010 by
PALGRAVE MACMILLAN®
in the United States—a division of St. Martin's Press LLC,
175 Fifth Avenue, New York, NY 10010.

Where this book is distributed in the UK, Europe and the rest of the world, this is by Palgrave Macmillan, a division of Macmillan Publishers Limited, registered in England, company number 785998, of Houndmills, Basingstoke, Hampshire RG21 6XS.

Palgrave Macmillan is the global academic imprint of the above companies and has companies and representatives throughout the world.

Palgrave® and Macmillan® are registered trademarks in the United States, the United Kingdom, Europe and other countries.

ISBN: 978–0–230–60799–6

Library of Congress Cataloging-in-Publication Data is available from the Library of Congress.

A catalogue record of the book is available from the British Library.

This book is printed on paper suitable for recycling and made from fully managed and sustained forest sources. Logging, pulping and manufacturing processes are expected to conform to the environmental regulations of the country of origin.

Design by Newgen Imaging Systems (P) Ltd., Chennai, India.

First edition: January 2010

10 9 8 7 6 5 4 3 2 1

This book is dedicated to Nicholas, Nathalia, and Leena, also Gavin, Catherine, and John Andrew

Also, in memory and remembrance of our colleague and comrade Joe L. Kincheloe December 14, 1950–December 19, 2008.

Contents

Foreword

Martha Montero-Sieburth

In the midst of the current economic worldwide crisis, global neoliberalism[1] is at the center of the storm. At no other time have the effects of neoliberalism in terms of its intent in global market liberalism, deregulation, and free-trade policies been more poignantly felt. The economic failure of Wall Street and the banking system has brought home the fragility of the credit and mortgage lending systems not only in the United States and Europe, but throughout the industrialized countries of the world.[2] Loss of remittances and work opportunities has already begun having a ripple effect in receiving countries throughout Latin America.

The economic safeguards that were created from the 1970s onward in capitalistic societies through neoliberalism with the promise of de-emphasizing government intervention in the domestic economy while focusing on implementing free market methods, freeing business operations, establishing property rights, and inducing privatization have seriously been compromised. The assumption that the unrestricted flow of capital would lead to the production of the greatest social, political, and economic good in any given society is now seriously being questioned.[3]

Adopted by the World Trade Organization, the World Bank, and the International Monetary Fund and the Inter-American Development Bank, some critics say the philosophy of neoliberalism has been imposed top-down to gain the "competitive advantage" favored by the opening of markets through government institutions and corporations.[4] This has lead to accusations from some critics that instead of development being fostered through these organizations and institutions, the setting up of economic power and domination of developed to nondeveloped countries has been promoted (Martinez and Garcia, 1996).

Those supporting neoliberalism consider the flow of capital as necessary for developing market efficiency, producing higher economic growth, and obtaining better returns on capital and investments—all processes that would eventually lead to development. Presumed was the idea that with such flows of capital resulting in efficiency, greater global stability would ensue, and more democratic forms of government and decision-making would develop. Yet from the current situation the world is in, even under the best of circumstances, the former outcomes have not realistically been achieved by many countries, and consequently the latter has not prevailed.

In this first decade of the twenty-first century, we have experienced instead extremely unstable global situations, with terrorism since September 11, 2001, becoming even more widespread. Train bombs exploded in Madrid on March 11, 2004, and in London in 2005. In September 2008, The ETA Basque separatist movement resumed bombings after having signed a peace ceasefire in 2006. The wars in Iraq and the incursions of Taliban fighters in Afghanistan have dramatically weakened the U.S. economy, and the continued saga between Israelis and Palestinians in the Middle East has diminished the possibility of sustaining bilateral peace agreements. Violence has become an accepted way of life and global destabilization is becoming more and more "normalized."

Concurrently, we are experiencing changes in our habitats, with global warming affecting not only humans but flora and fauna as well. Attempts at developing more democratic forms of government and decision-making processes, even in countries such as Bolivia and Venezuela that have initiated bottom up socialistic governments, have met with resistance and become mired in political haranguing, with platforms of ill-directed decision-making and subversion supplanting democratic tendencies.

In such a milieu, the economic bubble of prosperity that would lead to a liberalizing democracy has burst wide open, and in its aftermath, residual fear and hesitation to act are replacing hope. While pension and retirement funds are being reduced, the fate of future schooling and the pedagogies and humanistic values that support education and social justice are in a state of suspended animation. Without doubt, they will be used as hostages in the economic decisions that are to come and will be placed under close scrutiny as the budgets for educational benefits, including those promised by No Child Left Behind, will be reduced by the diminishing returns of the dollar and other currencies. Even with the election of President Obama, discussions of budget allocations for education continue to be the fodder of speculation on websites and blogs as reductions are expected.

Fear has overtaken our trust and belief in human nature and in the ensuing months the gains made by educational programs and projects created to "equalize" the playing field between affluent and poor school districts and between professionally competent teachers who are "highly qualified" and those who are not, as well as curriculum programs such as special education and Title I reading programs, will run the risk of being reduced, cut, or eliminated.

The processes being produced with the expansion of globalization will have further repercussions, and, as Nina Glick Schiller and Peggy Levitt (2006) point out, will affect cities as contexts, homeland politics, and the flows of capital, media, objects, and ideas as part of such globality. Gleaned from this current crisis is the obvious cause and effect that markets have on human lives, but more significantly, the fragility and susceptibility that education has in relation to the whims of the free market and the influences of a neoliberal ideology. Having been introduced since its inception into many of the schooling programs and practices worldwide, neoliberalism and the influences of globalization have become the mainstay of many schooling outcomes. Carlos Torres (2002) points out that "education within the nation-state . . . [has] been shaped by the demands . . . to prepare labor for participation in the economy and to prepare citizens to participate in the polity" (p. 363). He sees the purpose of schooling to be one of making labor highly skilled and competitive and education to be focused on problem-solving issues. Moreover, education has, in his opinion, become directed toward a marketplace rather than toward human rights ideology, and in such a process the democratic notions of citizenship that include tolerance, conviviality, and respect for human rights become secondary (Torres, 2002).

We have witnessed its effects in different countries and contexts, with my own country of origin, Mexico, becoming the personification of neoliberal policies and practices.[5] In some instances, issues of social justice have been sidestepped and ignored, as was the case in Chile, one of the most representative neoliberal success stories. In others they have been progressively infused with the idea that "what works" is what counts and that "best practices" are to be emulated worldwide, as in the case of the United States with the North American Free Trade Agreement and Britain. In other instances, issues of social justice have been commodified or, even worse, made redundant, as has been the case in Mexico. The introduction and development of the "maquiladoras," or export assembly plants, as part of the inroads of globalization has in effect changed the nature of Mexican family structures, the immigration flow of many Mexicans, and education.

Reverberations of Paulo Freire's teachings seem appropriate at this juncture, both in terms of his message, but also in terms of the healing power brought by his message.[6] In fact, as Peter McLaren (1999) has commented, Freire's critical pedagogy of possibilities serves to offer practical alternatives to the (neo)conservative and (neo)liberal discourses and practices. Of Freire's philosophy, several strands seem pertinent to the contextualization of this foreword: his use of language, notion of education, criticity, and historicity.

As we focus on the outcomes of education into the future, Freire's *Pedagogy of the Oppressed* (1974) early on depicted the influence of neoliberalism in education. The very issue of language containing structures of power from the landowner to the peasant is not that far fetched from our current discourses about corporate giants in relation to earners and taxpayers. Freire links education to human nature and moral order, making education an essential part of human life. In rereading Freire, Glass (2001) comments: "Language, culture, history and community are dependent on education, on freedom and the capacity to create forms ('ways') of life" (p. 17). Thus it is this triumvirate that is raised by Freire in his identifying voice and criticity. In a 1985 edited videotape Freire stated: "We should challenge the students concerning the right they have to have voice, the duty they have to be critical in having voice, and getting criticity by experiencing the voice. We cannot get the criticality without speaking." It is these three levels of understanding that remind us of what has been lost in our being seduced by a neoliberal stance that in fact has not contributed entirely to the "good" of the society, but to fulfilling the benefits of only a few. Such levels of understanding include: (1) the right to have voice, (2) the duty to be critical in having voice, and (3) becoming critical in experiencing voice. These are the basic tenets of a truly democratic society, yet through appropriation of a marketplace ideology that makes labor highly competitive, focus on problem-solving and not problematizing, and infusion of neutrality into the current national standards for curriculum, teacher professionalization, certification programs, and educational reform, these rights and duties become "silenced."

Equally important may be that education as a dialogical process so aptly described by Paulo Freire (1974) in this context may become less of the life of biology that he considered became life as biography, that is, life as history. In our enculturation process into social practices by groups and institutions defined around certain values and interests, the possibility of creating dialogical processes can only be engendered when we question the very groups and institutions by which we are socialized and can change the way that we interpret such a reality (Gee, 1988). Schools have a major charge in facilitating the way that biology can become history, but in the current state, the history that most likely will unfold and can be told will be that of the permanence of social inequalities and social injustices.

Counteracting what might be a pessimistic view is the publication of *Revolutionizing Pedagogy: Educating for Social Justice within and beyond Global Neoliberalism.* Not only is its timing propitious, but its messages most needed. Distributed throughout its twelve chapters are not only the frameworks that help the reader understand how critical pedagogies are being organized in different continents and in different contexts, and through different structural mechanisms like the testing industry, but also the strategies and practices that are possible for creating transformative change in education. This dual approach that fuels and unifies movements to grasp, and in that grasping begins to break and supersedes the ties of global neoliberalism helps us recover our meaning of critical pedagogy from the different discourses and positions presented.

Beginning with case studies from Chile to South Africa to the confrontation of corporate America in public education, these serve as frameworks to organize the prevalent discourses existent in schooling today. The included strategies for practicing critical pedagogy hone in on asking what education can do in Britain, United States, India, and the rest of the capitalist world. The work then explores positions that rethink Marx and Engels, examines schools that have been kidnapped, explicates strategies needed in teacher education and the understanding of knowledge-based economies, and posits striking lessons and ways of thinking about education.

The readings jolt us into the discourses of the moment, helping us question the chances education has to actually democratize and educate students. Woven through several of the chapters is Freire's notion of historicity, the insertion of self in the creation of history and culture (Montero-Sieburth, 1985). Ronald David Glass (2001) states: "The praxis that defines human existence is marked by this historicity, this dialectical interplay between the way in which history and culture make people even while people are making that very history and culture" (p. 16). It is in the struggle to be free that not only the possibility for humanization takes place, but also conversely, the possibility of dehumanization.

Creating history strikes a chord as the re-democratization of Chilean education meets the Penguin Resistance of the subtle changes being demanded by teachers, students, and community members head on. The use of community initiated research in the case of working class organizations in their ability to confront neoliberalism by identifying the delivery of social justice promises made in South Africa is another case of historicity. Confrontation of the corporatization of public education through the testing industry in the wake of No Child Left Behind heightens our awareness of the synergistic relationships that government, corporations, and the media use to control public education in the United States and leads us to question and develop strategies that inform the public about the privatization of schools, enable teacher education programs to integrate critical inquiry, allow students to learn about the historical developments around standards and assessment movements, and engage teachers not only in extensive research, but more importantly in becoming effective agents of change. Introduced are fresh ways of thinking about teachers and "the knowledge economy discourse." Several projects are advanced that modernize school, personalize learning, "scientize" teachers' knowledge, and focus on the "biologization and neurologization" of the learner and the commodification of schooling. In their realization, it is expected that they will develop a different education for teachers and learners and raise questions about the society and learner that is being configured.

Reading through each page compels us to face the contradictions and tensions inherent in pedagogical practices and to seek answers in the strategies that are described. Gleaned is an understanding of neoliberal and globalization influences in increasing greater homogeneity in societies through the power and operation of multinational corporations and free market ideology on the one hand, and on the other hand, as Carlos Torres (2002) indicates, greater heterogeneity and diversity, as environmental actions and democratization also increase. It is in creating this tension, as Freire (1985) has said, that the educational task cannot be understood. In this respect, *Revolutionizing Pedagogy* serves to expand our sense of pedagogy beyond our instructional practice and into the realms of knowledge production, cultural dissemination, and critique while simultaneously educating for social justice within and beyond global neoliberalism.

Notes

1. There are multiple definitions of neoliberalism showing a tension between whether it is viewed as a set of economic policies, a more commonly applied definition in Latin America, or whether it is viewed

as a philosophy, which is reflected in attitudes to the society, individual, and employment and is more commonly used in Western market democracies than in poor regions, according to Paul Treanor. (http://web.inter.nl.net/users/Paul.Treanor/neoliberalism.html)

2. Elizabeth Martinez and Arnoldo Garcia identify the following points for neoliberalism: (1) the rule of the market, (2) cutting public expenditure for social services, (3) deregulation, (4) privatization, and (5) eliminating the concept of "the public good" or "community" and replacing it with "individual responsibility." They cite a scholar who refers to "neoliberalism means the neo-colonization of Latin America" and refer to global neoliberalism as "the rapid globalization of the capitalist economy on a global scale." (http://www.corpwatch.org/article.php?id=376)
3. http://skeptically.org/wto/id10.html.
4. Ibid.
5. According to Martinez and Garcia, during the first year of NAFTA, the wages in Mexico declined 40–50 percent while the cost of living rose by 80 percent. "Over 20,000 small and medium businesses have failed, and more than 1,000 state-owned enterprises have been privatized in Mexico." (http://www.corpwatch.org/article.php?id=376)
6. Heinz-Peter Gerhardt (1993) points out that Freire philosophy, system, and generative themes have remained at the center of educational debates in critical pedagogy during the past three decades to the chagrin of many traditional First World academics.

References

Freire, P. (1974). *Pedagogy of the Oppressed.* New York, NY: The Seabury Press.

———. (1985). *Silence and Pedagogy.* Edited videotape in the repository of Martha Montero-Sieburth.

Gee, J.P. (1988). The legacies of literacy: from Plato to Freire though Harvey Graff. Review of the legacies of literacy: continuities and contradictions in Western culture and society. *Harvard Educational Review*, 58(2), 195–212.

Gerhardt, H.P. (1993). Paulo Freire. *Prospects: The Quarterly Review of Comparative Education* (pp. 439–458). Paris, FR: UNESCO, International Bureau of Education, (XXIII) ¾.

Glass, R.D. (2001). On Paulo Freire's philosophy of praxis and the foundations of liberation education. *Educational Researcher*, 30(2), 15–25.

Glick Schiller, N., and Levitt, P. (2006). Haven't we heard this somewhere before? A substantive review of transnational migration studies by way of a reply to Waldinger and Fitzgerald. Working Paper, 06–01. Center for Migration Development, Princeton University.

Martinez, Elizabeth, and Garcia, A. (1996). What is neoliberalism? A brief definition for activists. National Network for Immigrant and Refugee Rights. (http://www.corpwatch.org/article.php?=376)

McLaren, P. (1999). A pedagogy of possibility: reflecting upon Paulo Freire's politics of education. *Educational Researcher*, 28(2), 49–56.

Montero-Sieburth, M. (1985). A rationale for critical pedagogy. Review of the politics of education: culture, power, and liberation by Paulo Freire and translated by Donaldo Macedo. *Harvard Educational Review*, 55(4), 457–463.

Torres, C.A. (2002). Globalization, education and citizenship: solidarity versus markets? *American Educational Research Journal*, 39(2), 363–378.

Acknowledgments

We would like to express our deepest gratitude to all those who helped and encouraged us through the completion of this book. It would not have been possible without the contributions of our collaborators to whom we owe a great debt.

First, sincere and genuine thanks to Tony Green and Glenn Rikowski's series at Palgrave, but for their continued support and solidarity.

The idea of this book came as a result of conversations among the three editors, colleagues and collaborators Sheila Macrine, Peter McLaren, and Dave Hill.

We also want to extend our appreciation to Martha Montero, an early collaborator of Paulo Freire's in the early 1980s, for writing the foreword of this book. Her contributions to the translations of many of Paulo works give her a unique insight into the world of pedagogy.

Also, thanks to David Hursh for writing the afterword to this volume. His writing and insights continue to inspire and also challenge me to be better.

Special thanks must go to our research assistant and technical reviewer, Eddie Baldwin. We could not have managed without his keen eye and outstanding organizational skills. Thanks to Dean Allan De Fina and Dr. Deb Woo, at The Deborah Cannon Partridge Wolfe College of Education at NJCU, for their support.

Sincere thanks to the folks at Palgrave Macmillan for their help in getting this book off the ground. Julia Cohen, our editor, didn't let us slip. A special thanks to Samantha Hasey, editorial assistant at Palgrave Macmillan, for helping to get the book through production, and Maran for managing the copyediting process. Also, a special thanks to Nathalia Jaramillo for her helpful criticisms and insights.

Thanks to all of our students over the years.

SHEILA MACRINE, DAVE HILL, AND PETER MCLAREN

Introduction

Sheila Macrine, Peter McLaren, and Dave Hill

> *Paulo Freire was concerned with the number of persons who let themselves be deceived by neo-liberal slogans and so become submissive and apathetic when confronted with their former dreams. Paulo used a metaphor for this situation: "They have gone to the other side of the river!"*
>
> —Ana Maria (Nita) Araújo Freire

This volume arrives at a precipitous moment in the history of modern capitalism. Capitalism meltdown (or, more specifically, the epic housing crisis and the financial collapse), coupled with the election of the first biracial U.S. president, leaves little doubt over who will set the agenda in the current battle over public education, health care, and the judiciary system. Henry and Susan Giroux (2008) remark:

> Once upon a time a perceived bastion of liberal democracy, the social state is being recalled from exile, as the decades-long conservative campaign against the alleged abuses of "big government"—its euphemism for a form of governance that assumed a measure of responsibility for the education, health and general welfare of its citizens—has been widely discredited. Not only have the starving and drowning efforts of the Right been revealed in all their malicious cruelty, but government is about to have a Cinderella moment; it is about to become "cool," as Prince Charming-elect Barack Obama famously put it.

One would be profoundly naïve in minimizing the connection between the current state bailouts to the private corporations and the systematic enactment of defunding public education and even more naïve in underestimating the impact on the lives of the oppressed. The rapid intensification of neoliberal practices in education legitimized by political passwords such as "NCLB" has constricted the range of possible educational futures to those that support capital's bottom line. While critical educators are facing tough times, it is undeniable that they have been on the very front lines not only in challenging neoliberal predatory policies in education, but also advancing real alternatives. Once again critical scholars are raising their voices, and working at the interstices of the capitalist system (De Certeau, 2002), echoing the clarion call of George Counts (1978 [1932]) that schools can be instrumental in building a new social order.

This book brings together a group of leading international scholars to examine the paradoxical roles of schooling in reproducing and legitimizing large-scale structural inequalities along the axes of race, ethnicity, class, sexuality, and disability and, at the same time, to offer strategies for individual mobility. Through exploring the social relations of power, powerlessness, and hegemony, new social arrangements can be imagined, constructed, and challenged in education and social life in general. This book represents, in the main, what McLaren and Farahmandpur (2005, p. 44) call revolutionary citizenship, one that "work[s] toward a new type of democratic governance and redistribution of economic and political power that results in an oppositional form of globalization (...) a form of creative collective action centered on the devalorization of capital as a process of dealienation." Faced with the current financial meltdown, this has become more imperative.

Radical Pedagogy: A Perspective

Critical or radical pedagogy is an approach to understanding and shaping the school/society relationship from the perspective of the social relations of production within capitalist societies. It is also a practical approach to teaching, learning, and research that emphasizes teaching through critical dialogue and a dialectical analysis of everyday experience. In short, it is about teaching through praxis. Its approach is democratic, and its aim is to bring about social and economic equality and justice for all ethnic groups. It upholds the principles of and struggles for race, class, and gender equality. Practitioners include feminist educators, labor rights advocates, queer theorists, and Marxist humanists, among others. There are generally two streams. One stream is left-liberal and attempts to make capitalist society more "compassionate" and more democratic so that it better serves the interests of the poor and economically disenfranchised. The second is generally grounded in a critique of capitalist society (often through an engagement with the writings of Marx) and attempts to work toward a socialist society through social critique and nonviolent dissent. Both approaches eschew sectarian proselytizing and are committed to dialogue and debate over the meaning and purpose of democracy in capitalist and noncapitalist societies. Both have been part of the educational curriculum in colleges of education for the past two decades.

Quite recently, the international political economy was the toast of the global ruling class, and the bourgeoisie saw it as their biggest opportunity in decades to join their ranks. Freemarketeers have been given the New World Order's imprimatur to loot and exploit the planet's resources and to invest in global markets without restriction and with impunity. The menacing concomitant of capital's destructive juggernaut is the obliteration of any hope for civilization, let alone democracy. The working classes are taught to feel grateful for the maquiladoras and deregulated "special economic zones" that are now sprouting up in countries designated to provide the cheap labor and dumping grounds for pollution for the Western democracies. They are taught that socialism and communism are congenitally evil and can only lead to a totalitarian dictatorship. In short, capitalism and the legitimacy of private monopoly ownership has been naturalized as common sense. However, since capitalism involves an internal contradiction that produces a countervailing force that opposes any dominant force, the same rapid intensification has called into being a keen awareness among educators of that constriction. Awareness has, in turn, compelled educators to explore and develop new frameworks and strategies that recognize, question, critique, and sometimes break at least some of the ties binding them to capital's demands. These new frameworks and strategies have the potential to revolutionize pedagogy so that it is more effective in working within and against the contradictions of neoliberal capitalism in order to educate for social justice.

Following Morton and Zavarazadeh (1991), this book understands "pedagogy" not as commonsensical classroom practices or instructional methods as such, but as a political commitment, as an act of producing and disseminating knowledges in culture, a process of which classroom practices are only one instance. From this position, all discursive practices are pedagogical, in the sense that they challenge and propose a theory of reality—a world in which those discourses are "true." For the proposed world of these discourses to be "obviously" *true*, people need to be "instructed" so as to find them true. The goal of hegemonic pedagogical practices is to interpolate individuals within systems of intelligibility from which the reigning economic, political, and ideological social arrangements are deemed to be incontestably true. This is another way of saying that we conceive pedagogy to be a means, in ideological terms, of constructing and maintaining subjectivities that are necessary for reproducing existing social organizations.

The resistance and opposition to these dominant discourses and to the social order they legitimate also come from, as Freire (1970) would put it, a pedagogy of cultural action, specifically, the pedagogy of critique, which inquires into and exposes the historico-material conditions of possibility of any discourse. The pedagogy of critique, in particular the Marxist historical materialist critique used by many of the writers in this book, works to break the divisive and neoliberal understanding that class subjects have of themselves, of being so many "different" and "unique" individuals whose fundamental interests are "naturally" incommensurate and therefore unchangeable. The pedagogy of critique works to uncover the material conditions of the discourse(s) of neoliberalism that these discourses themselves attempt to hide. That is, the pedagogy of critique exposes economic (class) relations as determinant in the production of subjectivities under capitalism. In a dialogical move, the work of critique simultaneously provides access to concepts capable of explaining, rather than explaining away, contemporary contradictions in capitalism. Through this double move, the pedagogy of critique makes it possible to see that things do not have to be as they are, and in such a way, organizes people for change by reorganizing their theoretical frameworks (Morton and Zavarazadeh, 1991, p. vii). In this sense, the pedagogy of critique needs to be seen as a political terrain challenging not only the constant attacks on democracy, but also particular subsumed eugenic and genocidal dominant policies and practices such as English-only movement (Macedo, 2003 et al.), standardized tests, rankings, and so on.

With this aim in mind, the book begins by critiquing several key discourses that have the effect of blocking transformative change. Specifically, part one, "Frameworks for Organizing Pedagogy," critiques the discourse that naturalizes schools as autonomous and therefore neutral, thus legitimizing as neutral practices of teachers that, though they may be well-intended, ultimately serve the interests of the capitalist class in the ways that they maintain existing social arrangements (Ross & Gibson, 2007). It also examines, the ways in which the discourse of rights become reified, detached in the theoretical imaginary from the historico-material forces that enable it (Motalo & Vally, 2002).

This is accomplished by examining specific examples of the devastating effect of neoliberal governmental policies on educational systems across the globe. Beginning in South America, in chapter one, "A Wolf in Sheep's Clothing, or a Sheep in Wolf's Clothing: Resistance to

Educational Reform in Chile," Jill Pinkney Pastrana examines Chile, the first country in the world to adopt a broad series of neoliberal educational reforms. These reforms, put in place in the early 1980s during the Pinochet military dictatorship and the monetarist Chicago School economic and fiscal dictatorship, have left the country a legacy difficult to overcome. In the years following re-democratization, Chile has embarked on a series of education reform efforts notable for their reemphasis on the public sector. Though there are elements of these reforms that represent both neoliberal and progressive recommendations, there has been resistance to the *Reforma Educacional Chilena* by teachers, students, and local communities. In June of 2006, and again in 2007, massive protests by high school students that demanded improved public education for all took place. This "*revolución pengüina*," which occurred throughout the country, offers an intriguing venue from which to consider ongoing education reform and a reawakened Chilean civic engagement. Following the protests of 2007, the government has taken several unprecedented actions to revisit some of the more problematic neoliberal structures that have remained intact. Most notable among these changes is the abolition of the *LOCE (Ley Orgánica Constitutional de Educación).* The effects these changes will have on public education, specifically in terms of equity and access, remain to be seen.

Moving east, authors Salim Vally, Enver Motala, and Brian Ramadiro demonstrate in chapter two South Africa's proud history of resistance in and through education, "Education Rights, Education Policies, and Inequality in South Africa." Over the past few decades this resistance has generated popular epistemologies and pedagogies against racial capitalism, of which the "peoples' education movement," "worker education movement," and "popular adult and/or community education movement" are examples (see Cooper, 2002; Motala and Vally, 2002). This praxis has diminished relative to the erstwhile struggle against apartheid but still exists. It's center of gravity today has shifted to the new, independent social movements as they resist the impact of neoliberalism and increasing poverty and inequality in postapartheid South Africa. This chapter examines this resistance and specifically the role of research in enhancing the capacity of working class organizations to confront neoliberalism. It assesses the methodology of some mass-based community initiatives that challenge the new state to fulfill its social justice promises, and details the struggle to challenge the costs of education for working class communities, policy flaws, and the policy response of South Africa's state education departments. Specifically, the authors present three instances of the interaction between poverty, popular research, education, and resistance. Some of these initiatives use participatory action research as an epistemological challenge to the established and formal endeavors of the social sciences. They concur that "those who have been most systematically excluded, oppressed, or denied carry specifically revealing wisdom about the history, structure, consequences, and the fracture points in unjust social arrangements" (Cammarota and Fine, 2008) p. 215. The chapter concludes with a discussion of what these case studies suggest about educational transformation and the relation of socially engaged research to community activism in postapartheid South Africa.

In chapter three, "Taking on the Corporatization of Public Education: What Teacher Education Can Do," Pepi Leistyna returns us to the United States, focusing on the political economy of the American testing industry and the neoliberal policies that support it. The chapter examines the ownership, intent, and regulation of the private forces that produce, provide materials, prep sessions, and tutorials for, evaluate, report on, and profit from these tests. By looking at and analyzing this powerful sector and the old-boy networks therein, it is much easier to understand the synergy that exists among government, corporations, and the media, and how this force has been controlling public education in the United States to the detriment of its people. It is from this position of awareness that activists can better mobilize against these forces and teacher education programs can be restructured if they are really intent on democratizing schools, in order to better meet the needs of educators and students alike.

The last chapter in this section, chapter four, Revolutionary Critical Pedagogy: The Struggle against the Oppression of Neoliberalism—A Conversation with Peter McLaren," is a fascinating interview with Peter McLaren who explicates his notions on revolutionary critical pedagogy, as well as imagining a socialist alternative to neoliberalism.

Part two, "Strategies for Practicing the Pedagogy of Critique," develops the broader critiques of part one by theorizing and disseminating discourses and practices that hold promise and hope for substantive transformation, by focusing on specific sites and practices.

In chapter five, "Class, Capital, and Education in this Neoliberal and Neoconservative Period," Dave Hill examines capitalist teacher education from a critical and revolutionary socialist perspective. He asks, "What does education do in Capitalist Britain, United States, India, in the Capitalist world? What can Marxist educators do about it? How and where?" Then he asks where should we, as teachers in different sectors of the education state apparatuses, put our efforts. These efforts must be taken while recognizing the limitations but also the opportunities of our efforts as socialist and critical educators, as socialist teachers, as people who try to work as critical organic public transformative intellectuals. Of course it is not easy. The structural limitations on progressive and socialist action through the ideological and repressive apparatuses of the state, such as schooling and "the academy"/universities, that work for the most part on behalf of capital are considerable. Non-promotion, sidelining, denigration, even dismissals are common among socialist activist teachers. Lots of Left and union activists have been there (e.g., see Hill, 2007; Rikowski, 2002).

Next, Wayne Au steps back, takes a broad look at the school system from a Marxist perspective. In chapter six, "Defending Dialectics: Rethinking the Neo-Marxist Turn in Critical Education Theory," he responds to critiques of Marxist analyses of the relationship between schools and capitalist production as being overly economistic and deterministic; many critical educational theorists turned to conceptions of "relative autonomy" and individual agency and found the works of Althusser, Gramsci, and others giving rise to the field of neo-Marxist educational theory. He argues that many of these critiques rely on a fundamentally faulty understanding of Marx and Engels' original dialectical conceptions of the relationship between society (schools included) and capitalist production. Using historical documents written by Marx and Engels themselves as evidence, he contends that Marx and Engel's original analysis of the relationship between social structures and the economic base was nonlinear, nonmechanistic, and nondeterministic, and thus concludes that the neo-Marxist turn away from what has been labeled "traditional" or "orthodox" Marxist analyses of education was unnecessary and unwarranted.

In chapter seven, "Hijacking Public Schooling: The Epicenter of Neo-Radical Centrism," João M. Paraskeva examines how public schools have been kidnapped by what he calls neo-radical centrist policies. The chapter attempts to unveil some of the strategies that underpin the very marrow of current neo-radical centristic triumphalistic posture. To understand the lethal effects of such policies it is necessary to be aware of how particular key concepts and practices have been clearly and gradually twisted and perverted, positively hijacked from the public sphere and coined with an economic flavored materiality. It is also important to observe how the neo-radical centristic triumphalism is deeply enmeshed within the politics of the commonsense and the role that the media plays in building a particular commonsensical framework.

What is the result of this hijacking? In chapter eight, "Critical Teaching as the Counter-Hegemony to Neoliberalism," John Smyth argues that contemporary society is increasingly experiencing a dramatic loss of social connectedness or "social capital." This takes expression in schools in the increasing tendency, especially as reflected in school reforms and restructuring, of regarding teachers' work in technical terms. He argues that the way of turning this wider loss of social capital around is to regard teaching as a social practice in which there is greater emphasis on teaching for social responsibility, democracy, social justice, and civility. He gives examples of how this might be possible through a critical approach to teaching. He begins by arguing that the kind of society we have and the relationships between society and schools that exists have much to do with how society regards teachers and the construals it places on schools and the work of teaching. Concomitantly, how teachers perceive their work and whether they are prepared to accept the dominant view of their work or instead resist it (Helsby, 1999; McLaren, 1998; Smyth and Shacklock, 1998) hold important implications for schooling as well as society in general. Pursuing this line inevitably draws him into the debate about non-subject-matter outcomes of schooling that was portrayed insightfully in the Elementary School Journal in May 1999 (Good, 1999). Finally, in this chapter, Smyth attempts to extend and broaden the nature of the discussions that must occur on this topic in respect of, and for the sake of, schools.

In chapter nine, "Empowering Education: Freire, Cynicism and a Pedagogy of Action," Richard Van Heertum recognizes that an understanding of such critiques of the current order of things is a necessary starting point for any progressive or radical pedagogy. Without an analysis of what is wrong, it is hard to convince anyone that change is necessary. And yet critique alone rarely inspires students to act. They need something to fight for together. The chapter thus looks at the centrality of hope as a necessary complement to critique in progressive education. Combining insights from Freire and Marcuse, he argues that critical pedagogies need to seriously engage methods to create an environment amenable to resistance and struggle for change, founded on a belief that a better world is possible. Freire (1994) offers inspiration and practical tools for the progressive educator that can be instrumental in realizing these goals. Marcuse offers two additional ingredients, a program of aesthetic education and a model for teachers as public intellectuals, combining utopian dreaming and embodied hope. Van Heertum thus argues that progressive pedagogy must move beyond consciousness-raising and identity formation alone to foster the belief that change is possible.

Finally, in chapter ten, "Teachers Matter... Don't They?: Placing Teachers and Their Work in the Global Knowledge Economy," Susan L. Robertson examines a number of projects underway that, if realized, have the capacity to generate profound changes to education and to teachers' work, to implement many of the ideas of the earlier mentioned critics. At the heart of these projects of "translating" the knowledge economy discourse into new institutional structures and material practices is the view that the education system must be radically transformed to secure the future. These projects include the "modernization of the school," "personalization" of learning, the "scientization" of teachers' knowledge, the "biologization/neurologization" of the learner, and the commodification of schooling. Robertson writes that "If I am right in my prognosis, then these developments can be read as a rupture in the grammar of schooling." In combination, these projects, if realized, would lay the groundwork for a very different kind of "education laboring" for teachers and learners. They also raise fundamental questions about what kind of learner and what kind of society is being constituted. The chapter is developed in three parts. Robertson begins with some reflections on the globalization of neoliberalism throughout the 1980s and 1990s and its consequences for teachers' work. Then she turns to an examination of the knowledge economy discourses promoted particularly by the international agencies from the late 1990s. This master narrative has gained sufficient traction in state policy circles for it to legitimize a newer, deeper round of institutional innovation/transformation, including education. In the third part of the chapter she briefly examines four projects intended to advance this renovation and recalibration of education to constitute a knowledge-based economy. In the conclusion, Robertson stands back and reviews the implications of these developments for contemporary societies more generally, and for teachers as laboring class in particular.

The Brazilian educator and political activist Paulo Freire (1996 [1970]) argued that while there are exceptional academics and a handful of organizations dedicated to conducting research that serves egalitarian ends, not enough academics are working as critical "cultural workers" who orient themselves toward concrete struggles in the public and political domains in order to extend the equality, liberty, and justice they defend (Freire, 1998). He maintained that "[t]he movements outside are where more people who dream of social change are gathering," but points out that "there exists a degree of reserve on the part of academics in particular, to penetrate the media, participate in policy debates, or to permeate policy-making bodies" (Shor and Freire, 1987, p 131).

Antonio Gramsci wrote extensively about the role of intellectuals, especially concerning their potential, as transformative intellectuals, in developing "good sense," a critical analysis of society, as differentiated from "common sense," somewhat akin to "folklore." He wrote that educators have, or can have, a privileged function in society. In this sense, as Gramsci put it, "[a]ll men [sic] are intellectuals . . . but not all men have in society the function of intellectuals," maintaining that the notion of intellectuals as being a distinct social category independent of class was a myth (Gramsci, 1971, p. 10). By "organic," Gramsci was referring to those intellectuals—such as school teachers—who remain or become part of the social class they are working with, linked to/being part of/participating in local struggles, campaigns, and issues, and who work toward developing critical consciousness—class consciousness and analysis. Many teachers and educators do this—taking on the issues and experiences and demands of the local community from which their students are drawn. This is in contrast to those "traditional intellectuals" who see themselves as a class above and separate from class struggles, or who, as knowledge workers (Kelsh and Hill, 2006) either ally themselves with the dominant (pro-capitalist) ideology, or critique neoliberal capitalism but show no sign of wishing to replace it.

The Task of Radical Educators/Teachers

Adherents of critical revolutionary pedagogy, critical revolutionary pedagogy, that is, radical educators and teachers who are democratic socialist or Marxist, seeks equality—far more equality of outcome in schools, universities, job rewards, pensions, longevity, life chances—rather than simply equal opportunities to get on in a capitalist society that remains inevitably unequal (Greaves, Hill, and Maisuria, 2007). As Dave Hill, one of the editors of this volume, suggests,

> The task of democratic Marxist/Socialist teachers, and of resistant egalitarian Socialist and Marxist counter-hegemonic and teachers, students, cultural workers, policy makers and activists is:
>
> 1. to expose and organise and teach against the actual violence by the capitalist state and class against the "raced" and gendered) working class;
> 2. to expose the ways in which they perpetuate and reproduce their power, that of their class, through the ideological and repressive apparatuses of the state (such as the media, the schooling, further education and university systems);

3. in particular the way they do this through demeaning and deriding the "cultural capital" and knowledges of the ("raced" and gendered) working class through what Pierre Bourdieu termed "cultural arbitrary" and "symbolic violence"—the way working class kids are largely taught they are worth less—or worthless—and upper class kids are taught they will control and inherit the earth, and some middle class kids are taught how to manage it for them;
4. argue for, propagate, organize, agitate for and implement democratic egalitarian change and policy. (Hill, 2009a; see also Hill, 2009b and Hill and Boxley, 2007)

Teachers and schools can use, can creatively develop, or "subvert" the formal and the hidden curricula to develop deep critical reflection in pupils and students concerning capitalist exploitation in the economy and society, and related and linked forms of oppression that ensure that the rich stay rich and the poor get poorer. And at how schools and universities—and the billionaire-owned mass-media—make such grotesque inequalities seem "only natural." At a wider level, radical teachers, teachers promoting equality, can themselves work toward a wide-ranging comprehensive restructuring of the schooling and education systems so that they could maximise both equality of opportunity and a far greater degree of equality of outcome, and also so that the negative labeling of millions of children in our primary schools, with its "raced" and gendered and other types of social-class-based stereotyping, discrimination, hurt, and inequality, can be replaced by a loving, egalitarian, and emancipatory system of schools, classroom experiences, and, ultimately, society. The work of critical pedagogy—and, in particular, the role of critical revolutionary pedagogy—and its exponents in universities, colleges, and schools, and our writing and teaching through books such as this can have a transformative role in this emancipatory and liberatory project.

References

Cammarota, J., and Fine, M. (2008). *Revolutionizing Education: Youth Participatory Action Research in Motion*. New York: Routledge.

Cooper, Frederick. (2002). Decolonizing Situations: The Rise, Fall, and Rise of Colonial Studies, 1951–2001. *French Politics, Culture and Society*, 20(2), 47.

Counts, G.S. (1978 [1932]). *Dare the School Build a New Social Order?* Carbondale, IL: Southern Illinois University Press.

De Certeau, Michel. (2002). *The Practice of Everyday Life*. Translated by Steven Rendall. Los Angeles, CA: University of California Press.

Freire, Paulo. (1994). *Pedagogy of Hope*. New York: Continuum.

Freire, Paulo. (1996 [1970]). *Pedagogy of the Oppressed*. New York: Continuum.

Freire, Paulo. (1998). *Teachers as Cultural Workers*. Boulder, CO: Westview Press.

Good, Thomas, L. (1999). Editorial Statement. *The Elementary School Journal*, 100(1).

Giroux, Henry A. and Giroux, Susan. (2008). *Beyond Bailouts: On the Politics of Education After Neoliberalism*. TRUTHOUT, Wednesday, December 31, 2008. http://www.truthout.org.

Gramsci, Antonio. (1971). *Selections from the Prison Notebooks* (p. 10). London: Lawrence and Wishart.

Greaves, N., Hill, D., and Maisuria, A. (2007). Embourgeoisment, Immiseration, Commodification—Marxism Revisited: A Critique of Education in Capitalist Systems. *Journal for Critical education Policy Studies*, 5(1). Online at http://www.jceps.com/index.php?pageID=article&articleID=83.

Hill, D. (2009a). Theorising Politics and the Curriculum: Understanding and Addressing Inequalities through Critical Pedagogy and Critical Policy Analysis. In D. Hill and L. Helavaara Robertson (eds.). *Equality in the Primary School: Promoting Good Practice across the Curriculum* (pp. 232–245). London: Continuum.

———. (2009b). Critical Pedagogy, Revolutionary Critical Pedagogy and Socialist Education. In S. Macrine, P. McLaren, and D. Hill (eds.). *Critical Pedagogy: Theory and Praxis* (117–139). London: Routledge.

Hill, D. and Boxley, S. (2007). Critical Teacher Education for Economic, Environmental and Social Justice: An Ecosocialist Manifesto. *Journal for Critical education Policy Studies*, 5(2). Online at http://www.jceps.com/index.php?pageID=article&articleID=96.

Kelsh, D. and Hill, D. (2006). The Culturalization of Class and the Occluding of Class Consciousness: The Knowledge Industry in/of Education. *Journal for Critical Education Policy Studies*, 4(1). Online at http://www.jceps.com/index.php?pageID=article&articleID=59.

Macedo, Donaldo, Gounari, Panayota, and Dendrinos, Bessie. (2003). *The Hegemony of English*. Boulder, CO: Paradigm Publishers.

McLaren, P., and Farahmandpur, R. (2005). *Teaching against Global Capitalism and the New Imperialism: A Critical Pedagogy*. Lanham, Md.: Rowman & Littlefield.

Morton, D. and Zavarazadeh, M. (1991). *Theory/Pedagogy/Politics: Texts for Change*. Chicago, Ill.: University of Illinois Press.

Motala, S., and Vally, S. (2002). People's education: From people's power to Tirisano. (2002). In Peter Kallaway (ed.). *The History of Education under Apartheid, 1948–1994: The Doors of Learning and Culture Shall be Opened*. South Africa: Pearson. Online at http://books.google.com/books?id=9OdroQajgHsC&pg=PA174&dq=motala+and+vally&source=gbs_toc_r&cad=7

Rikowski, G. (2002). Fuel for the Living Fire: Labour-Power! In A. Dinerstein and M. Neary (eds.). *The Labour Debate: An Investigation into the Theory and Reality of Capitalist Work*. Aldershot, UK: Ashgate.

Ross, E., Wayne Ross, and Gibson, Rich, eds. (2007). *Neoliberalism and education reform*. Creskill, NJ: Hampton Press.

Shor, I. and Freire, P. (1987). *A Pedagogy for Liberation: Dialogues on Transforming Education*. South Hadley, MA: Bergin & Garvey.

Smyth, John, and Geoffrey Shacklock. (1998). Re-making teaching: ideology, policy, and practice. *Issue 4 of Occasional paper* (p. 1.) London: Routledge.

Part I

Frameworks for Organizing Pedagogy

Chapter 1

A Wolf in Sheep's Clothing or a Sheep in Wolf's Clothing: Resistance to Educational Reform in Chile

Jill Pinkney Pastrana

Introduction

This chapter is part of a body of work grounded in ongoing ethnographic research begun in 1998 that examines the local effects of education reform that has been occurring in Chile since re-democratization. Chile is a country with a long history of national commitment to a well-developed public education system with levels of school coverage and national literacy significantly higher than those in most countries in Latin America. In the early 1980s while in the midst of a brutal military dictatorship, Chile became the first country in the world to re-form its educational system along firmly neoliberal lines. Since re-democratization, a series of educational reforms have been undertaken that, while progressive on some levels, have not significantly changed the underlying neoliberal structure put into place in the 1980s. These reforms, many of which have been housed within *La Reforma Educacional Chilena* (REC), have met with some success, though many problems remain. The REC and other reforms brought a new impetus to address the many problems in Chilean schooling, especially in the public sector, and created many opportunities for schools throughout the country to innovate and improve. However, the reform did not address in any way the underlying neoliberal structure that was imposed upon the educational system of Chile during the dictatorship, and, in fact, some elements of the REC were actually designed to improve neoliberal aspects of the educational system.

An approach toward an analysis of this post-dictatorship reform effort can begin by examining two key questions:

1. Was the purpose of the *Reforma Educacional* to address the disintegration of the education sector following the neoliberal changes of the 1980s?
2. Was the *Reforma Educacional* an attempt to "fine-tune" the neoliberal changes that had been made?

Interestingly, an examination of the data suggests that both of these questions can be answered in the affirmative. This chapter will revisit these questions and consider the data with regard to the reactions and resistance from several key actors who have been particularly impacted by education policy, namely, the teachers and students in Chile. These are the people on the front lines—Chilean students and education professionals who struggle every day with a deteriorating infrastructure, low social status, and extremely challenging learning and working conditions. Meanwhile, teachers, students, and their communities struggle to maintain dignity within a traditionally authoritarian system bent on forming students capable of high test scores and unquestioned capitalist consumption.

Within this backdrop, an interesting reaction to the reforms has been playing itself out. It is apparently contradictory on several levels. The Chilean government has been systematically increasing the investment in education since re-democratization in 1989. The preparation, pay, and working conditions of teachers have improved, especially in many of the country's municipal schools. The content and scope of the curriculum has been enriched, and educational establishments even have the opportunity to create locally responsive curriculum and programs that open up vast possibilities to democratically serve marginalized communities. Yet, perhaps because of the traumas of past reforms, distrust of state institutions, and limitations that still remain locked into the structure of the system, the full potential of the REC has not been met and continues to be greeted with resistance on many levels.

What is the nature of this resistance? From teachers? From students and community? Does resistance to government reform represent an understanding and critique of the underlying neoliberal structure upon which all but perhaps the most recent reforms have been placed? Hence, does resistance represent a refusal to accept any reform before the neoliberal underpinnings are eliminated? This orientation echoes the position of the "*la izquierda no reformada*," the Leftist elements in Chile, who have both individually and in terms of political representation rejected formal participation in government as long as the Constitution implemented by the military regime remains in place. It is a resistance to government claims to change (even if the change might be for the better) because of the realization that dictatorship policy remains largely intact and the goals of the reform may not really be in the interest of the parties involved. Have the reforms since re-democratization really been in essence a "wolf in sheep's clothing"? Or, are the reforms misperceived by a public that remains skeptical of anything that reeks of the legacy of the Pinochet era where undoubtedly problematic policies have left the public education sector in a state of destruction and disarray. As a "sheep in wolf's clothing"—current, albeit incremental, reforms may be a positive pathway to democratic change, yet resistance "by all means necessary" and the skepticism toward the possibility of "using the master's tools to dismantle his own house" create a complex challenge for Chilean education policy makers.

Legacy of the Dictatorship—Fallout from Neoliberal Reform

In order to more clearly consider the current situation of education reform and resistance to reform in Chile, it is helpful to contextualize the structural and ideological factors that have played a role in the political economy of Chile's education system in recent years. I will begin by briefly contextualizing the state of Chilean schools as found during the years of 1998–99. This information was gathered as a part of a sixteen-month ethnographic research project in the ninth region of Chile. It is not meant to be specifically representative of the conditions in *all* Chilean schools, yet when examined in light of other regional and national data, this more focused, ethnographic data does capture many of the challenges lived daily by teachers and students throughout the country. The period in which these data were initially gathered was nineteen years since the initial reforms of the dictatorship were put into place, ten years since "re-democratization" and one year into the REC, which was formally set into motion in 1997. It is important to note that in all the years since re-democratization, education reform has continually been underway. The REC elevated education reform to national prominence and placed the debate of education firmly into the political arena.

Schools in Chile—Choice and *Subvenciones*

Sabemos que están aquí porque somos los peores y no saben que hacer con nosotros. Este escuela es clasista y racista. Si tuviéramos apellidos europeos estaríamos en los primeros cursos.—Sophomore "problem" student.

[We know we are here because we are the worst and they don't know what to do with us. This school is classist and racist. If we had European last names we would be in the highest courses.]

—Pinkney Pastrana, 2000

The Chilean school system is characterized as falling into four separately funded types: the *municipal*, *particular subvencionada*, *particular pagada*, and *corporaciones* (municipal/public, semi-private, private, and corporate run). Concurrently with the decentralization of the system in 1980, the military government created the system of *subvenciones,* a national voucher system. Ideologically, education policy and the condition of schooling in Chile today are similar in many respects to that found in the United States. Indeed, virtually all reform efforts taking place in the United States today with No Child Left Behind mirror the path taken in Chile thirty years ago. In the arguments put forth for vouchers, charter schools, calls for national standards and testing, and merit pay for teachers, the neoliberal agenda is clear. Even in the opposition to NCLB in the United States and the recent reform efforts in Chile, the means, measures, and outcomes of education reform being proposed reflect nothing less than the hegemony of neoliberal and neoconservative ideology founded in a distrust of government and public sector organizations and buoyed by utopic appeals for policies enabling privatization, competition, and other free market approaches to schooling.

The vouchers, or *subvenciones,* in Chile are a per-student payment given directly to each establishment. Only the fully private schools do not receive these *subvenciones.* There is a general tendency to believe municipal/public schools to be the very lowest, least desirable form of education. The years of dictatorship and neoliberal education policy, for all its petty complexity reduces primarily to cuts in funding, have completely eroded the physical infrastructure and professional morale within public schools. Not to mention the erosion of trust in the competency of this once strong and enviable public system. Since the plebiscite of 1989 effectively restored "democracy" to Chile, the *Consertación* government has slowly been restoring the funding levels in education, though it had not, until perhaps the most recent changes of 2007, changed any of the fundamental tenets of dictatorship education policy, namely, the structures for funding education, the role of the state in education, the organization within regions and municipalities overseeing the functioning of schools, and the national and local provisions that have enabled the proliferation of privatization and drained resources from the public sector.

In Chile, as elsewhere, the markers of social class fall along certain lines, some economic, others ethnic and cultural, and still others racial. This strict social stratification is clearly evident in Chilean schools, marked by blatant and virtually uncontested tracking. Tracking is especially evident at the high school level and between the different types of high schools or *liceos*. These two types of institutions, *humanístico-científico* and *técnico-profescional*, Scientific-Humanist (SH) and Technical-Professional (TP), respectively, tend to serve very different populations of students. The SH schools are generally more oriented toward university preparation while the TP schools offer students the opportunity to graduate with a specific title, indicating preparation for differing types of jobs, such as auto mechanics, tourism, food preparation, carpentry, and so on.[1] It is widely assumed that students in the TP schools are those of lower academic ability, not really college material. There is overwhelming evidence that obtaining access to higher education in Chile is a highly classed, raced, and gendered affair, though recent data suggests that Chile enjoys an impressive social mobility index among all sectors with the exception of the very top tier of income (Spilerman and Torche, 2004). In Chile, classism is alive and well, and students who wish to advocate for more educational opportunities may be accused of "*Buscando un futuro que no le corresponde/*looking for a future that doesn't fit them" (Pinkney Pastrana, 2000).

Examining table 1.1, there can be no doubt of the discrimination and lack of equity and opportunity in terms of rural and urban student populations. At the time of this research there was not a single elite private school that offered vocational education. There is not a single elite private school in a rural area. There is not even a public municipal school that offers rigorous academic preuniversity preparation in a rural area. High school students in the rural areas of the ninth region can only look forward to a future of menial work. The possibility exists that they take the route chosen by some families to leave the countryside and live as *internados*, boarders, in schools in the urban areas, usually miles from family support.[2] This choice entails not only the separation from family and often undermines the means of survival in families who rely on young members to participate in agricultural labor, but also usually requires an additional financial commitment that many rural families simply cannot meet.

Table 1.1 Number of establishments by geographic region and school type, ninth region, 1997

School type	Geographic location	Scientific humanist		Technical professional	
		High school	Adult ed.	High school	Adult ed.
Municipal	Urban	37	17	24	1
	Rural	0	1	1	0
Semi-private	Urban	20	6	26	0
	Rural	0	0	6	0
Private	Urban	10	2	0	0
	Rural	0	0	0	0
Corporate owned	Urban	0	0	6	0
	Rural	0	0	0	0

Source: Departamento de Estadística. División de Planificación y Presupuesto Compendio de Información Estadística, Ministerio de Educación (1997).

Though a strong contingent of municipal schools still pursue the SH curriculum, a considerable number of these have considered or are in the process of developing programs in accordance with the demands of changing their status to that of a TP institution. For much of the working class, university education is an expensive unknown, while a technical degree offered by TP schools is often seen as providing a more practical education in terms of future employability.

Sociologist James Coleman began this discussion in the 1960s in his extensive study of U.S. public schools. He concluded that achievement was more strongly linked to students' socioeconomic than any other factor. The higher the social class of a student, the higher that student's achievement (Rothstein, 2004). Though direct links between teachers' beliefs of student ability and a child's social economic status are readily evident, as is also the case in the United States, discrimination on this basis is usually justified through arguments of student ability and motivation. On visiting schools and classrooms throughout the region, one is hard pressed to find working class students in higher-level courses in any school. Since neoliberal policies were implemented in the 1980s, the potential of public schools to serve as a space for mixing students representing multiple class backgrounds has greatly diminished. The tendency for educational access to rest on a family's ability to purchase a satisfactory education in the free market reverberates through the various levels of education in Chile. Today, very few working class students from public schools ever set foot in a university.

Erosion of Community

One of the consequences rarely discussed in arguments calling for the decentralization of schools and vouchers is the erosion of community that occurs when families feel that quality schooling is found outside their immediate neighborhood. This reality presents an interesting contradiction in terms of the usual marriage between debates calling for decentralized school practices and open enrollment: "local control" and "choice." Many argue that the only effective instantiation of decentralized schools results when a highly organized, local constituent has pushed for control of its local schools (Guari, 1998; Parry, 1997). Charter schools in the United States sometimes fulfill these criteria (Wells et al., 1996a,b). In these cases, local communities seek to develop educational opportunities and specific curricula not offered in mainstream public schools. These schools may become centers of the local community, and have the potential to create unique and powerful opportunities for underserved populations. Up until recently, this was not the case in Chile. Even today, there are very few schools throughout the country that take advantage of the possibilities offered them by the de-centralized nature of the system to incorporate "alternative" social visions into their curricula.[3] Decentralization of education in Chile appears to privilege decentralized funding rather than emphasizing the potential that "local control" can hold for innovative curriculum. This has led to a devastation of schools in economically marginal communities, while not inspiring creativity or innovation and schools that support diversity in perspectives.

Table 1.2 helps to give us an idea of the monetary and temporal cost de-centralization has exacted on the Chilean population. We can easily make several interesting observations regarding the effects of de-centralization on community schools in Chile. In most communities, at least one school lies within walking distance, yet parents have chosen to send their children to schools within eleven–twenty-five blocks of their homes. This loss of time due to travel and its cost is a major stress on many families. Still, we see the faith in education as a social promoter held by many families, as evidenced by their willingness to further extend this sacrifice. We also see evidence of the loss of confidence in the public municipal schools in Chile. The general assumption throughout the country is that any school is better than a municipal school. In fact, this is not always the case. Another factor revealed by table 1.2 is the enormous differences in terms of liquid assets and monetary investment held by families who make up the upper-income and lower-income in Chile. The use of an automobile to provide transportation to and from school reveals the deep breach between the "haves and the have nots" so characteristic of societies that have adopted neoliberal economic policy (Boron and Torres, 1996; Petras and Veltmeyer, 1999; Samoff, 1996).

The examples examined in table 1.2 also illustrate the legacy of classism, historic amnesia, and authoritarianism that define so much of the social reality in Chile today. The social reproduction that reenforces the classist, racist, and sexist status quo continues virtually unchallenged in schools across Chile today despite the advent of the REC and re-democratization. The traditional practices and authoritarianism found in schools have exacerbated the difficulty of encouraging both school communities and others to create spaces where participation in democratic processes and public dialogue is the norm.

The absence, until relatively recently, of public spaces and democratic discourse underlies the realities found in Chilean schools, the university, and society at large and has created a cultural logic resistant

Table 1.2 Household expenditures and transportation per student by school sector

School sector	Municipal *n* = 490	Semi-private *n* = 520	Private *n* = 181
Total payments to school in 1993 (in pesos)	6,682	23,626	624,330
As a fraction of household income	0.007	0.015	0.096
Weekly cost of transportation to and from school (in pesos)	233	470	2,335
Number of blocks to school	13.7	14.7	25.2
Time spent traveling to and from school (in minutes)	13.6	15.4	18.8
Student travels by car (%)	1.2	6.7	51.4
Student walks to school (%)	70.0	54.6	14.4
Pesos per month family is willing to spend if child could attend one of the five best schools in the metropolitan area	8,226 (*n* = 420)	13,427 (*n* = 474)	65,146 (*n* = 152)
As fraction of household income	0.079	0.085	0.122
Travel time family would accept if child could attend one of the five best schools in the metropolitan area (in minutes)	42.0 (*n* = 457)	40.5 (*n* = 487)	28.2 (*n* = 163)

Source: Guari, 1998, p. 120.

to democratizing processes. This resistance was born out of the years of oppression during the Military Regime, and is coupled with a highly political "apolitical" sarcasm that critiques the failure of post-dictatorship, neoliberal Chile to fulfill so many of its promises. Throughout the months of this study, I was often struck by the frustrating limitations to potentially progressive change caused by traditional conceptualizations of "proper" professional activities and the boundaries defining them. Teachers, both in the universities and schools, students, parents, and administrators maintained their narrow definitions of professional and social involvement. They often became uncomfortable and resistant when asked to breach traditional boundaries in favor of "inter-disciplinary," collaborative, or open-ended activities. A principal from an elementary school in a small *comuna* commented, "We are so frustrated by the Ministry, they always come up with one reform or project after another, and now they are leaving the projects open for us to design. They should just give it to us prepared, tell us what to do" (Pinkney Pastrana, 2000). This comment clearly demonstrates the clash between traditional practices in education systems (reenforced by neoliberal structures supporting standards and testing) and the need for new cultural practices of schooling that can exploit the potential for progressive reform. Here resistance takes the form of opting out of the process in favor of the "path of least resistance."

Touting "local control" as a model for dealing with different sociocultural needs and preferences remains problematic. In advocating for a broad-based public school mandate of a national curriculum that includes issues of diversity or other potentially controversial topics, a national, centralized system can guarantee that, at least to some degree, school policy practices and curriculum will not be hostage to the arbitrary will of local communities. It is very difficult in a completely decentralized system to institutionalize an effective and rigorous process for setting content standards and school policies and procedures.[4] Thus, at this point, schools in Chile continue to reflect a conservative legacy and they remain at the mercy of the particular biases of the local school bosses, the municipal mayors (*alcaldes*). In terms of innovative curricular offerings, schools tend to follow the dictates of the market, emphasizing high-status language programs, for example (such as schools that offer extra English, French, or German language instruction), or low-status but popular programs offering technical degrees found in the TP schools.

Teachers

> *La clase es muy tradicional, el profesor dicta mucho y los alumnos parecen estar de acuerdo con esto. Él tiene buena relación con sus alumnos, es muy permisivo, yo diría que en extremo, por lo cual los estudiantes le tienen mucha confianza, ya que también bromea con todas sus alumnas. Sin embargo, me han informado, que el relajo de este profesor se debe a que trabaja en tres Liceos, por lo cual llega muy cansado a clases y trata de pasarlo bien con sus alumnos.*
>
> *[The class is very traditional, the teacher mainly lectures and the students seem content with this. He has a good relationship with his students, he is very permissive, perhaps to the extreme, and the students appear to be very friendly with him, the female students all joke around with him. However, the teacher told me that his attitude is due to the fact that he works at three different high schools, and he arrives to class very tired and just tries to have a good time with his students.]*
>
> —Pinkney Pastrana, 2000

In monthly workshops with about seventy teacher-collaborators throughout the research period, as well as in workshops held with fifty administrators participating in Ministry of Education professional development, two striking consistencies indicating the social and material conditions of the educational realities in Chile became almost immediately apparent. First, teachers as well as administrators always indicated their lack of time. Second, teachers and administrators generally felt in the midst of some sort of "cultural crisis" in which the cultural deficits of students and their families are primarily to blame.

The problem of time is a direct consequence of the changes made in the educational system in the 1980s. "Free Market" logic sought to instill a spirit of competition and efficiency in schools through decentralization and privatization. Part of the mechanism to attain this efficiency was changing the traditional teacher benefits, despite their already limited status, through pay cuts, flexible contracts, and the annihilation of the possibility of collective bargaining. The profession of teaching is now a very low status occupation in Chile. Teaching, even in higher income areas, is a profession characterized by harsh working conditions and very low pay. In a neoliberal education system where test scores determine access to each major field of study, education requires the lowest scores on University entrance exams. In surveying a class of fourth-year teacher education students, only four out of twenty-eight indicated they wanted to be teachers. The remaining students were studying education because it was the only field their test scores qualified them to enter.

Perhaps more than anything else, it is the low pay of teachers that is responsible for their lack of time. Though the REC and other reforms since re-democratization have begun to improve the situation, many teachers are still paid by pedagogical hour. Though municipal contracts usually allow greater time for activities outside of the classroom (this varies depending on the municipality), there are virtually no controls for the fiscal decisions made by semi-private and private schools. Teachers are usually not compensated for time needed to prepare for classes, correct work, or work collaboratively with fellow professionals. In order to earn a living wage many teachers seek work in several schools because teachers may not work in more than one school of a given type. In other words, they cannot hold more than one position in a municipal, semi-private, or private school simultaneously. These "taxi teachers" rush from one school to another in an effort to earn a living. Several participants in this study worked in excess of fifty pedagogical hours—fifty forty-five-minute classes per week. One colleague in the Universidad de la Frontera's *Programa de Innovación a la Formación de Profesores* worked seventy pedagogical hours per week. The dilemma is obvious. How are teachers to carry out reform innovations: creating constructivist collaboratively designed lessons, giving alternative and thoughtful feedback, creating spaces where students can "learn to learn," and even reconceptualizing themselves as respected professionals within society under such conditions?

One of the primary goals of the REC's *Jornada completa* is that teachers commit themselves to serving only one establishment full-time. This is in order to address the problem of time and facilitate building a professional community with fewer obstacles to working collaboratively. It is striking that except for the government's plan to gradually increase the minimum teacher pay (which is included as a minor part of the reform, falling under the rubric of *El Desarrollo Professional de los Docentes*), the reform fails to place the issue of teacher remuneration at the fore. The National Ministry makes broad-based decisions concerning teacher salaries, yet municipal governments still have control over the ultimate commitment to educational funding within any given municipality. It is the problem with teacher pay that, perhaps more than anything else, continues to inspire resistance from teachers and undermine the success and potential of education reform.

Rather than concentrating on the national need for drastic increases in teacher pay, the plans put forward by the REC titillate competitive desires—not educational innovation—through monetary and professional "goodies," such as merit pay and lengthy fieldtrips to foreign countries to observe other's pedagogical practices. Of course, Chile has a long history of advancement through "*pituto*," social connections. It truly is a case of "not what you know but who you know." These bonus awards aimed at rewarding promising professionals and outstanding programs often end up rewarding the faithful and obedient supporters of reform-funded programs, rather than searching the broad and talented body of Chilean education professionals.

Issues of "cultural crisis" often referred to by teachers and university students alike were curious for several reasons. Teachers and administrators are generally caught between feeling in the midst of some sort of "cultural crisis" and feeling like the "successful," hyper-developed, modern, "free-market miracle" they are often touted to be. These contradictory impressions inspire a situation of dialectical dissonance. Social discourse has been forced to disarticulate alternative options to social and economic development, to forget historical ruptures in Chile's political and economic development, to deny the hardships and failings of the economic development of the present and content itself with the trappings of "success" as defined by "McMall" global capitalism. In terms of the student realities that inevitably make their way into the classrooms, this "crisis" is sometimes articulated in terms of social conditions. For example, the real challenges associated with living in poverty are recognized as primary factors responsible for the difficult situation students and their families must face, thus making their receptivity to good pedagogical practices almost impossible. More often, however, when pedagogical practice is questioned and problematized, revealing, for example, traditional, authoritarian, or inequitable practices, such as the blatant tracking system found within Chilean schools, the students and their supposed lack of family support were blamed for difficulties. The notion that "children do not fail in schools, schools fail the children" is rarely part of the educational conversation. This cultural deficit assumption of students, especially rural and poor students in municipal schools, rages almost uncontested among teachers and preservice education students alike and crosses political boundaries. Left-leaning university students display as harsh a belief in the capabilities, or lack thereof, of students coming from differing classes and cultural backgrounds as do right-leaning students. The reform has thus far failed to open up a dialogue or begin the debate on cultural or class identities. Though this may be slowly changing, there has been little debate concerning the role of the Chilean education system in reproducing the status quo.

With the exception of the *Colegio de Profesores* (the de facto teachers union), and until recent student protests demanded national attention, the last thirty years have seen a shift in the culture of education from that of a conservative, though politically active, traditional bureaucracy to one of a largely silent, often fearful and suspicious, self-censoring, often apolitical constituency. Resistance, opposition, and discontent have been often expressed through frustrated reactions directed at students and among colleagues. Where formal education reform has lagged behind in concretely addressing the issue of marginalization and deficit ideology in education, the recent mobilization of high school students has placed the consequences of such ideology front and center in their critique of education in Chile. The student protests of 2006 and 2007 vociferously and dramatically demanded both structural and ideological critiques of neoliberal policies, and their aftermath in Chile.

La Reforma Educacional Chilena

> *La educación chilena está cambiando con ritmos y sentidos que no tienen precedentes históricos. Chile va adquiriendo conciencia que sin renovar su sistema educativo y sin cambiar sus formas de enseñar y aprender no podrá ir más lejos en su desarrollo. La renovación de la educación ha pasado al primer lugar de la agenda pública. Por su parte, las familias priorizan altamente la inversión en educación, muchas veces sacrificando otras opciones. El gasto público y privado en ella está experimentando un importante incremento, no obstante que todavía es insuficiente para cubrir necesidades y demandas que se hacen cada vez más exigentes.*
>
> *La sociedad chilena discute ampliamente sobre la educación que quiere. Hay grandes consensos, pero también disensos. Sobre esta base, la educación desarrolla dinámicas promisorias.*
>
> *[Education in Chile is changing with a rhythm and sense without historical precedent. Chile is becoming conscious that without a renovation of their education system and without a change in the manner of teaching and learning, it will not be able to advance in its development. The renovation of education has become the first priority in the public agenda. For their part, families highly prioritize the investment in education, often at the sacrifice of other options. The public and private cost of this is experiencing an important increase, but this amount is still insufficient to cover the necessities and demands that are becoming every moment more demanding.*
>
> *Chilean society broadly discusses what education it wants. There is grand consensus as well as disagreements. On top of this foundation, education develops dynamic promisors.]*
>
> —José Arellano, Introduction to *La Reforma en Marcha*, MECE 1998

In 1997 Chile embarked on an ambitious and far-reaching education reform initiative: *La Reforma Educacional Chilena.* Noteworthy on many levels, this reform represented an interesting approach to Chile's education challenges. The REC, while promoting many very "progressive" elements in education, represents a strong reinvestment in the public sector. At least at the "surface level," the REC represents a departure from policies put in place during the 1980s in Chile. This reinvestment in the public sector and more "humane" approach toward equity in educational opportunity is emblematic of education policy post re-democratization. However, the REC simultaneously seeks to reenforce some of the blatantly neoliberal structural elements of the Chilean educational system that were put in place during the years of the military dictatorship.

The REC has been a massive effort to improve education in Chile. There have been various changes in the REC since its inception in 1997, but many of the goals, specific programs, and approaches to education reform continue to this day. There are four primary areas through which specific reform programs are organized:

1. *Jornada Completa*: Projects that support the infrastructure and training that will facilitate a full school day in all Chilean municipal schools.
2. *Aprender de Aprender*: Curricular reform, renovation, and decentralization of elementary and secondary curriculum.
3. *MECE (Mejoramiento de la Calidad de la Educación)*: This focus has been the largest part of reforms in recent years. MECE programs, which cover the spectrum from preschool through university level education, provide a means through which some of the reinvestment into public education can take place. Development of innovative practices, technology, encouragement of professional initiatives and the creative capacity of teachers are all specifically targeted by MECE projects.
4. *Desarrollo Profesional de los Docentes*: Focused on teacher development and training. Support and new programs and grant opportunities that target initial teacher training, professional development of current practitioners, scholarships for international professional development, and continued education, National Awards for Excellence in Teaching.

The specifics described in this list are all oriented within a set of three primary foci that provide an overarching framework for recent and ongoing educational reform in Chile. These three areas that also serve to define much of the ideological bent of post-dictatorship reform in Chile are "*Calidad/*Quality," "*Equidad/*Equity," and "*Participación/* Participation." In an analysis of these terms, we find several striking and seemingly contradictory elements.

"Quality," as a focus in education reform without being specifically defined, is a meaningless word. No politician, regardless of their ideological or party persuasion, would argue against promoting quality education in their country, or state, or region. However, the path that leads to this quality education, and how precisely quality education is defined and measured, varies widely. These definitions are heavily laden with ideological biases to what represents the most efficacious path toward "excellence," how excellence is defined, and ultimately which actors within our communities benefit or are excluded by the policies that follow the clarification of these definitions.

"*Equidad*" is a concept that represents a significant change in education policy in Chile. Considered representative of "socialist" ideals during the dictatorship, highlighting ideals of equity in education forces the state to directly confront the reality of inequity in education that exists in Chile today. As we have seen in the previous discussion, the fallout from neoliberal policies implemented during the dictatorship left broad educational devastation in Chile, felt especially dramatically among the more marginal sectors of society. In an environment in which the Government Social Report of 1984–85 stated; "Values that correspond to models of unattainable lives should not be fomented among the poorer sectors of society"(ODEPLAN, 1985), it is not surprising that discussions concerning educational equity were nonexistent. Today this has changed. The discussions are varied, some leading in directions that may not realistically be helpful for traditionally underserved communities, others representing innovative, progressive, and even transformational potential. These discussions continue as education reform responds to the political dynamic of the local and the global.

It is within the framework of "participation" that the contradictory ideological positions, at once progressive, and at once neoliberal, underlying the educational structure in Chile can be unearthed. Within the various Ministry of Education reform programs that seek to improve the participation of various actors detailed previously, virtually all contain this potential contradiction. Each of the reform agendas are seeking to aid reinvestment in public education, and improve the access and quality of schooling available to all sectors of the population, but they are proceeding within a clearly neoliberal structure that has not been changed from its initial implementation in the early 1980s. Beyond and within the structural, funding, and assessment realities that remain unchanged since the dictatorship flourishes the ideological reality of free-market thinking. During the dictatorship, the "logic of the market" did become embedded in the logic of schooling. This neoliberal ideology permeates everything and has led to an absence of critique within Chile (save within a few sectors) and, until recently, a paucity of possible alternatives to free market education policies.

Critiques of REC—*Revolución Pengüina*

In June of 2006 and again in 2007, resistance to education reforms and the stasis of a dysfunctional system prompted an impressive and inspiring broad, national, and largely youth-led series of protests that succeeded in shutting down the country and illuminating the serious problems in public education in Chile in a way reminiscent of the civic activism that prompted the plebiscite and the end of the military regime. These protests, named the *revolución pengüina*, in reference to the school uniforms worn by high school students in Chile's public schools, revealed a massive and well-articulated critique of schooling in Chile. The critique of the *pengüinos* is in many ways similar to the critique previously articulated by the *Colegio de Profesores* (*Colegio de Profesores*, 1999). The mobilization of the *pengüinos* is dramatic, powerfully calling attention to the failures of the education system. In many ways this explosion of youth activism appears to signal a new and growing sense of civic engagement. Until the *pengüinos* shut down the country, broad democratic participation, except in small and rather focused instances, had remained lackluster in post-dictatorship Chile. The *pengüinos* have ushered in a new period of youth activism and engagement, and this movement may signal an opening of the public sphere in which demands can be made for alternatives to the hegemony of neoliberalism.

The *revolución pengüina* specifically targets the "wolf" lurking in the shadows of incremental change. These protests appear to be a blanket rejection of lingering neoliberal logic underlying even the seemingly "progressive" elements of recent education reforms. Interestingly enough, the directive of the protests has consisted of students from various political parties. Some student leaders represent various leftist ideological positions, and student representation from youth leaders associated with the *UDI* and *Renovación Nacional* parties (traditionally far-right-wing parties) have been active in articulating the demands of the students as well. It is clear that students from across the political spectrum recognize the "education crises" in which Chile finds itself (Gonzales, 2006). It may be the case that the various student voices, as articulated by the leaders of the protests, are not unified in their critique of the education system in Chile. It may also be the case that the critique of Chile's high school students is less ideological in nature and focuses on surface level problems rather than on structural impediments to democratic change. Yet, a solid critique of neoliberalism has emerged most vociferously articulated throughout the various phases of student protest. It is compelling to consider the ramifications that a critique of neoliberal policy coming from youth representing both the right and the left may hold for the future of politics and policy in Chile.

On December 28, 2007, the student leadership of the *pengüinos* presented their proposal for reform to the minister of education. The student proposal reveals an interesting breadth of focus regarding many elements of the educational system. Their discussion included a variety of proposals notable for their democratic suggestions (*Congreso Segundario Nacional,* 2007).

- Funding
- Extension to the full school day/*Jornada escolar completa*
- Organization of schooling (*Ciclos*)
- Teaching
- School leadership teams
- TP education
- University entrance exams/ *Prueba de Selección Universitatia (PSU)*
- National student card
- Residential school students
- Curricular adjustments

All of the proposals embedded in the foci are very compelling to consider. The specific proposals overwhelmingly constitute a critique of and challenge to many of the previous policies from the neoliberal reforms of the dictatorship years. Perhaps the most blatant rejection of neoliberal policies is the proposals related to funding and educational organization. The student leadership in Chile has developed a very pointed critique about just how "local control" in education funding, administration, and municipalization was designed, its purposes as well as the fallout of these policies.

> En la práctica, la aplicación de este modelo de municipalización y su perpetuación por los gobiernos de la Concertación significó el

> establecimiento de una "*educación para pobres*." Una fabricación de mano de obra barata no calificada cuyo bajo costo jugará un rol funcional al proyecto capitalista neoclásico imperante. Todo esto bajo la bandera ideológica de la libertad de enseñanza.[5] (*Conclusiones Generales* Congreso Nacional de Dirigentes Secundarios, 2007)
>
> In reality, the application of this model of municipalization (decentralization) and its perpetuation by the Concertación governments signifies the establishment of an "education for the poor." A manufacturing of cheap, untrained labor whose low cost plays a functional role in the prevailing neoclassical capitalist project. All of this under the ideological banner of "freedom in education." (*Conclusiones Generales* Congreso Nacional de Dirigentes Secundarios, 2007)

Given this perception, the students have proposed an end to *subvenciones* and both the *municipal* and *particular-subvencionado*/semi-private types of schools. In a *subvencion* free system, it is proposed that funding for schools and programs be administered through the regional ministries and not remain in the hands of local schools under the jurisdiction of municipal leadership/mayors. In essence, this plan restores the fiscal responsibility for education to the state. In a similar vein, it is proposed that the state, via regional ministries, has a reinvigorated power to administer curricular and organizational standards to Chilean schools. As discussed earlier, schools (and semi-private institutions in particular) previously had relative independence across a broad spectrum of administrative and curricular decisions. This proposal invites the state to again take the leadership role in organizing and guaranteeing high standards and quality schools across the country.

Several of the highlights of the student proposals are noteworthy in their focus on re-democratizing educational sites, and enhancing the equitable distribution of educational services throughout the population. Proposals concerning school leadership teams contain suggested changes to the composition of these teams, and stand in stark contrast to the former, more authoritarian school leadership structure. Previously, the school director, perhaps with some input from the parent organization, was responsible for issues brought to the attention of these organizations. The *pengüinos* propose broadening the representation of this group to students, student (class) representatives, representatives from ethnic organizations, representation from the *colegio de profesores*, representatives from the unclassified staff, and parents. These changes would significantly broaden the level of representation to include many more key demographics that participate in the institution of schooling on a daily basis. A similar democratic vein is seen in the proposals for free student transportation, more government funding for better support to students in TP schools (especially when students do their practical experience), increased electives offered in schools for all students, and some limitations on standardized testing. Proposals also call for higher teacher standards, higher salaries, training, professional development, and random evaluation of all teachers every semester—evaluations that should include student input.

Another interesting shift to democratic and solidarity issues includes the suggestion to add curricular content concerning "ethnic cultures," "ethnic language" (these being dependant on and responsive to local minority languages and cultures), and "civic education." The student proposals even take into consideration the living conditions of students who live as boarders at some schools. These students often come from rural areas with limited educational opportunities. The student proposal is to change the nature of living conditions to be more community based and friendly. The bulk of student reform proposals reveal a strong tendency toward a more democratic organization of schooling. This seems to reflect a growing democratic culture of civic engagement in Chile as it continues to rebuild its democratic participatory culture following the decimation of democracy during seventeen years of dictatorship.

Current State of Affairs—Response to *Pengüinos*

On June 13, 2007, in response to the mobilizations of the *pengüinos*, the government of president Bachelete and the Ministry of Education responded with a number of very interesting reforms that suggest a potential dramatic change in the organization of education in Chile. It has begun with nothing less then the elimination of the *LOCE*. In its place a new "law," the *Ley General de Educación* (LGE), has been established. This most infamous educational legacy of the dictatorship, the *LOCE,* was initiated on the eve of the re-institution of democracy in Chile. It guaranteed that the strong arm of the dictatorship would extend into the democratic government and maintain the free market educational structure it had established. The primary policies that had begun in earlier years that were reinforced with the *LOCE* were: (i) Decree # 3,063, which began the municipalization of K-12 schools, and (ii) Decree #3,476, which created government subsidies to private and public schools for each student enrolled, creating a "voucher" system for schools level K-12. So the *LOCE* further institutionalized "local control" and vouchers in Chile.

With the *LOCE* technically no longer in existence does this mean that funding and other organizational issues can at last be resolved? The answer appears to be yes and no. A new law, the LGE, has been put into place. A brief signed by the president states that the LGE:

> es más inclusivo, transparente y participativo, pues reconoce el derecho de los padres a ser actores relevantes del proceso educativo de sus hijos, además del derecho a organizarse y exigir educación de calidad a los establecimientos durante todo el proceso de formación académica. (MINEDUC, 2007)
>
> [...is more inclusive, transparent, and participatory, it recognizes the rights of parents to be relevant actors in the educational process of their children, also the right [of the state?] to organize and demand quality education from educational establishments throughout the entire academic process.]

The LGE directs that the state become the principal guarantor of quality education in Chile—not local constituencies, beginning with anti-discriminatory mandates to all educational institutions and a commitment to keep all institutions and actors accountable. It also promotes transparency in the right of all actors to participate in the school counsel and a new energy toward democratization in the Chilean education system. All of these proposals thus far seem to precisely match the demands made by the *pengüinos*; things are looking promising. Additionally, the LGE will be setting new requirements for all establishments to meet high standards. However, one final piece of the LGE could well reduce all of the positives gained with the ministry's response to the students' protests; this one element alone stands in stark contrast to the demands of the *pengüinos* and the progress that seems to have been made. The ministry will be *increasing* the amount of the *subvenciones*. The voucher system will remain in place and be, in fact, strengthened by the LGE. How these newest changes will play out in Chilean schools remains to be seen.

Again the question can be asked: Is the purpose of these most recent reforms to significantly change the neoliberal structural foundation underpinning the educational system in Chile? Or rather, is it a matter of massaging the system in order that the neoliberal system can function more freely and efficiently? Are the reforms really a "wolf in sheep's clothing or a sheep in wolf's clothing"? As with the REC the implications and goals of the LGE are unclear. Perhaps a stronger ideological argument needs to be developed regarding the ultimate goals of and commitment to education in Chile. The education system cannot simultaneously serve the marginalized populations in Chile and disenfranchise populations through market measures that favor the "haves" over the "have nots."

It will be interesting to observe exactly how much "market logic" must be eliminated in order for equitable and excellent public education to develop. Chile has said "*adios general*" but it appears to be not quite ready to say "*adios carnaval*."[6] Chilean education policy makers would do well to heed the words of one of their most senior scholars Ivan Nuñez, former minister of education during the Allende years. When asked about changes in education policy in Chile from 1980 onward, Nuñez (1999) replied, "I think the neo-liberal policy makers did us a favor, now we know what definitely does not work in education."

Notes

All translations appearing in this document are the author's unless otherwise noted.

1. In the course of this research I became aware of several unpublished studies that had found troubling evidence that TP schools further marginalize students by having a very low rate of actual degrees awarded. The rate of graduation within the TP schools of students possessing a technical degree is less than 20 percent throughout the country. Often students complete the course of study but fail to complete their final projects, often entailing payments of special project fees or hidden costs of project development, and may never be awarded their technical degree. Leaving with a very mediocre academic preparation, these students are unable to work in the industrial, blue-collar sector for lack of a *título,* nor are they able to apply for any higher level university course of study.
2. In 1997, the ninth region registered 17,937 students living as boarders. Of this total, 9,128 attended municipal schools and 8,529 attended semi-private institutions. Only 7 of the total number attended private institutions (Compendio de Información Estadística, 1997.
3. One of the rare examples is the *liceo* in Chol Chol. Run by the Catholic Church, this school's mission is to provide a native-language educational option for the region's main ethnic minority—the Mapuche. The school has explored the potential offered by de-centralization to create its own unique program. It holds bilingual classes and has developed innovative curricula using the largely rural and indigenous realities that define the lives of its students.
4. Former regional minister of education (IX Region) Guillermo Williamson related one particularly troubling instance. One of the schools in the ninth region refused to allow a female student to enroll

due to the fact that she was pregnant and unmarried. Though this clearly represents an affront to the rights of young women, as well as the overwhelmingly oppressive and conservative ideas held by many traditionalists in schools, there were no legal means available to the Regional Ministry to require that the school respect the civil rights of this student. The final decision was up to the *alcalde* (mayor), who supported the position of the school administration. The student and her family had no recourse in the matter.

5. In reality, the application of this model of municipalization (decentralization) and its perpetuation by the Concertación governments signifies the establishment of an "education for the poor." A manufacturing of cheap, untrained labor whose low cost plays a functional role in the prevailing neoclassical capitalist project. All of this under the ideological banner of "freedom in education."
6. Lyric taken from a song, "Adios General," by the Chilean rock group "Sol y Lluvia."

Bibliography

Aedo-Richmond, R., and Richmond, M. (1996). Recent curriculum change in post-Pinochet Chile. *Compare: A Journal of Comparative Education*, 26(2), 249–247.

Arenas, A. (2004). Privatization and vouchers in Columbia and Chile. *International Review of Education*, 50(3–4), July 2004, 379–395.

Austin, R. (2003). The *State, Literacy, and Popular Education in Chile, 1964–1990*. Maryland: Lexington Books.

Bonnefoy, P. (2006). Chile looks (slightly) left. *The Nation*. http://www.thenation.com/directory/bios/pascale_bonnefoy.

Boron, A., & Torres, C. A. (1996). The impact of neoliberal restructuring on education and poverty in Latin America. *Alberta Journal of Educational Research*, 42(2), 102–114.

Brunner, J.J. (2005). Comparative research and public policy: from authoritarian to democracy. *Peabody Journal of Education*, 80(1), 100–106.

Carnoy, M. (1995). Is school privatization the answer? *Education Week*, 14(40), 52–60.

———. (1998). National voucher plans in Chile and Sweden: did privatization reforms make for better education? *Comparative Education Review*, 42(3), August 1998, 309–337.

Colegio de Profesores de Chile. (1999). *Propuesta Educacional de Magisterio. A la ciudadanía y a los candidates a la Presdencia de la República*. Santiago, Chile.

Compendio de Información Estadística. (1997). Ministerio de Educación, División de planificación y presupuesto. Compendio de Información Estadística, Santiago, Chile.

Congreso Nacional de Dirigentes Secundarios. (2007). *Conclusiones Generales*. Mejillones, Región de Anto Fagasta, 2007.

Congreso Secondario Nacional. (2007). Propuesta final año 2007: Sintisis general. Available online in the Portal Pingüino: http:// www. portalpinguino.cl/organizate/propuesta-para-educacion-de-los-estudiantes-secundarios-de-chile/.

Cox, C. (1998). "Reforma curricular y el rol de Educación en el nuevo milenio." Professional Development Workshop. April 4, 1998. Chile, Temuco: Universidad de la Frontera.

De Laire, Fernando. (2002). El discurso del 21 de mayo y los debates emergentes. *Revista Mensaje*, July 2002, 20–23.

Economist. (2006). "How to make them better," 381(8498), 3–5.

Gardner, W. (2005). Choice cheats children. *American School Board Journal*. July, 2005. http://www.asbj.com/MainMenuCategory/Archive/2005/July.aspx.

Gonzales R.P. (2006). Dirigentes secundarios: el perfil de los líderes. *Universia Crónica*. http://www.universia.cl/portada/actualidad/noticia_actualidad.jsp?noticia=10798.

Guari, V. (1998). *School Choice in Chile: Two Decades of Educational Reform*. Pittsburgh, PA: University of Pittsburgh Press.

Magendzo, A. (2005). Pedagogy of human rights education: a Latin American perspective. *Intercultural Education*, 16(2), 137–143.

Martin, T. (2003). Divergent ontologies with converging conclusions: a case study comparison of comparative methodologies. *Comparative Education*, 39(1), February 2003, 105–117.

McEwan, P. (2001). The effectiveness of public, Catholic, and non-religious private schools in Chile's voucher system. *Education Economics*, 9(2), August 2001, 103–128.

Ministerio de Educación, Chile (MINEDUC). (2007). *Presidenta de la República Firma Proyecto que Reemplaza la LOCE*, April 9, 2007. Available online: http://www.mineduc.cl/index.php?id_portal=1&id_seccion=10&id_contenido=4825.

Narodowski, M., and Nores, M. (2002). Socio-economic segregation with (without) competitive education policies: a comparative analysis of Argentina and Chile. *Comparative Education*, 38(4), November 1, 2002, 429–451.

Nuñez, I. (1999). Personal Interview, Ministry of Education. Santiago, Chile.

ODEPLAN. (1985). Encuesta de caracterización socioeconómica de la población. Santiago: CASEN, *Informe Social* 1984–1985, 21.

Parry, T.R. (1997). Decentralization and privatization: education policy in Chile. *Journal of Public Policy*, 17(I): 107–133.

Petras, J., and Levia, F.I. (1994). *Democracy and Poverty in Chile: The Limits to Electoral Politics*. Boulder, Co.: Westview Press.

Petras, J., and Veltmeyer, H. (2005). Empire with Imperialism: The Globalizing Dynamics of Neoliberal Capitalism. London: Zed Books.

Pinkney Pastrana, J. (2000). *Subtle Tortures of the Neo-Liberal Age: The Case of Chile*. Unpublished dissertation.

———. (2007). Subtle tortures of the neo-liberal age: teachers, students, and the political economy of schooling in Chile. *Journal for Critical*

Education Policy Studies, 5(2), November 2007, http://www.jceps.com/?pageID=article&articleID=102.

Puiggrós, A. (1996). World Bank education policy: Market liberalism meets ideological conservatism. *NACLA Report on the Americas XXIX*, (6): 26–31.

Rothstein, R. (2004). *Class and School: Using Social, Economic, and Educational Reform to Close the Black-White Achievement Gap*. New York City: Teachers College Press.

Samoff, J. (1996). Chaos and certainty in development, *World Development*, 24(4), 611–633. [re: "financial-intellectual complex"]

Schiefelbein, E. (2004). The politics of decentralization in Latin America. *International Review of Education*, 50, 363–383.

Spilerman, S., and Torche, F. (2004). Living standard potential and the transmission of advantage in Chile. In E. Wolff (ed.). *What Has Happened to the Quality of Life in the Advanced Industrial Nations* (chapter 8, pp. 214–253). Northampton, MA: Edward Elgar.

Teitelboim, V. (1988). Fascistas, filósofos y lectores. *Araucaria de Chile* (41): 49–57.

UNESCO. (1995). *Everybody's Business*, 66. http://unesdoc.unesco.org/images/0015/001502/150262e.pdf.

Valverde, G. (2004). Curriculum convergence in Chile: The global and local context of reforms in curriculum policy. *Comparative Education Review*, 48(2), 174–201.

Wells, A.S. et al. (1996a). "Globalization and educational change." In A. Hargreaves et al. (eds.). *International Handbook of Educational Change* (Vol. 3, pp. 3–17).

Wells, A.S. et al. (1996b). "Underlying policy assumptions of charter school reform: the multiple meanings of a movement." Paper presented at the Annual meeting of the American Educational Research Association, New Orleans.

Chapter 2

Education Rights, Education Policies, and Inequality in South Africa

Salim Vally, Enver Motala, and Brian Ramadiro

This chapter examines the resistance and specifically the role of research in enhancing the capacity of working class organizations to confront neoliberalism. It will assess the methodology of some mass-based community initiatives that challenge the new state to fulfill its social justice promises. The chapter specifically details the struggle to challenge the costs of education for working class communities, policy flaws and the policy response of South Africa's state education departments.

We present three instances of the interaction between poverty, popular research, education, and resistance. Some of these initiatives use participatory action research as an epistemological challenge to the established and formal endeavors of the social sciences. We concur that "those who have been most systematically excluded, oppressed, or denied carry specifically revealing wisdom about the history, structure, consequences, and fracture points in unjust social arrangements" (Cammarota and Fine, 2008, p. 215).

The chapter concludes with a discussion of what these case studies suggest about educational transformation and the relation of socially engaged research to community activism in postapartheid South Africa.

Introduction

Over the last few years, socially engaged researchers and activists in South Africa have participated in a range of research projects in education and other areas of social policy. They have deliberately conceptualized and conducted research in and through communities where the challenges of poverty, oppressive conditions, and social exclusion are pervasive. In doing so, they have explicitly stimulated discussion about these conditions and their underlying causes to mobilize responses and raise public consciousness about the issues confronting such communities. This has also led to the emergence of stronger links between socially committed researchers based in academic institutions and community activists. It has stimulated the development and uses of research in communities, having direct effects on the accountability of public representatives on the use of public resources and by implication on how academic scholarship is viewed.

A number of reasoned arguments have been advanced to support such a perspective about the development of research. Perhaps the most important of these is the realization that often social research "objectifies" communities and reduces complex theoretical and practical issues to data gathering and often produces outcomes that are patronizing and even offensive to such communities. This is largely because, especially in regard to communities that are bedeviled by egregious poverty and its social effects, research regards such communities as being in "deficit," devoid of history, knowledge, thinking, and struggle. The solutions to "their problems" lie, in the perspective of such research, in some external intervention based on the advice of researchers and consultants. The expertise of the latter is relied upon to make "recommendations" about the "problem." Often this means the intervention of researchers and consultants with little or no orientation to the deeper social characteristics of such communities, their class composition, history, language, culture, traditions, and experience of resistance, and even less to their reservoir of knowledge and experience and social consciousness.

These arguments are not to be conflated with arguments about local approaches to knowledge, as we argue later, but point to the critical importance of historical and contextual approaches to social analysis and for a better understanding of the character of such communities. They argue the importance of transcending the limitations inherent in superficial survey data and the case for richer and complex qualitative data that moreover recognize the categories of gender, social class, "race," and identity necessary for meaningful social analysis.

While the surveys referred to in this chapter do not pretend to resolve the complexities of research, they are nevertheless important illustrations of the attributes of such research and its importance to historical and contextual analysis. They demonstrate both the context within which particular social questions arise and assert the importance of adducing qualitative information as *complementary* to the value of data adduced through surveys and its associated methods. Such a complementary approach is important for the production of evidentiary data that has both qualitative and quantitative attributes having value both to social analysis and practice. Not only is such analysis important to enhance theory and the explanation of social phenomena, but it is also instructive for developing resistance to the effects of social policy and practice that perpetuate poverty and exploitative and oppressive social conditions. Such approaches to social enquiry based on solidly grounded information and knowledge can also have value for engaging with the policy and decision-making agencies of the state and with public representatives. Primarily though, the research relies on the assumption that it has value for the mobilization of strategies and for planning and organizing of local education and other campaigns to both inform local action and democratize issues relevant to educational struggles in communities.

A further assumption of such research is that in order to resist the power of dominant discourses and the practices based on these, communities would need to form autonomous organizations representative of their interests and seek greater public accountability from state agencies and decision makers. In South Africa, despite the promise of postapartheid reform and some positive gains made through it, democratic accountability through open engagement about how public choices are made remains a serious problem. At this time the almost continuous cycle of sometimes violent public protest action taking place across the country, arising from issues of "service delivery," is, despite the denials of some bureaucrats and politicians, a concrete expression of the failure of such accountability and democratic engagement.

The lack of service delivery around education, housing, health (particularly the HIV/AIDS pandemic), electricity, sanitation, and water has once again made townships and informal settlements into "hotbeds of activism"; the minister of safety and security recorded fifty-eight hundred citizen protests in 2005 (Bond, 2006). Out of these sustained protests, mass organizations such as the Landless People's Movement, the Anti-Privatisation Forum, Anti-Eviction Campaign, the Treatment Action Campaign, and the Abahlali Base Mjondolo (shack-dwellers' movement) have been strengthened. Many of the community protests are spontaneous and do not always lead to sustainable organizations and social movements. In recent times, these protests have not abated. In a written reply to a question posed in the National Assembly, the minister of safety and security conceded that there were over ten thousand protests in the years 2006/2007 (National Assembly, 2007). The new social movements have increasingly allied themselves with local education resistance. The praxis of South Africa's ERP, discussed later in this chapter, is based on the struggles of these community organizations.

Out of these sustained protests, mass organizations such as the Landless People's Movement, the Anti-Privatisation Forum, Anti-Eviction Campaign, the Treatment Action Campaign, and the Abahlali Base Mjondolo have arisen. These new social movements have increasingly allied themselves with local education resistance. The praxis of South Africa's Education Rights Project, discussed later in this chapter, is based on the struggles of these community organizations as well as on teacher unions, student organizations, and working class parent bodies.

Many of the new social movements use the tools of research to "systematically increase that stock of knowledge which they consider most vital to their survival as human beings and to their claims as citizens" (Appadurai, 2006, p. 168). In doing so they "de-parochialize" research by breaking the monopoly of experts and specialists on the means of knowledge production and create the conditions for social movements to engage in the "capacity to aspire"—linking individual problems to the larger social, economic, and political forces and endeavoring to transform these (p. 176). This also entails a democratic commitment to knowledge and those for whom social research should be undertaken.

The Promise of "Transformation": The Historical Context

The year 1994 was indeed an important year for the people of South Africa. It signaled the demise of the hated apartheid system and the commencement of a process for the establishment of a democratic state that was to be secured through the development of a constitution in which basic civil and socioeconomic rights would be entrenched. The South African Schools Act (DoE, 1996), following the Constitutional right to basic education, defined basic education as ten years of compulsory schooling for all learners. This act was an important beginning and signaled the intention to make education a key social goal. A number of other reforms were also signaled both through the raft of policies that were passed to "transform" the education system and through a series of practical administrative and other arrangements affecting many aspects of the educational system. For instance, the racist and fragmentary system of apartheid administration was supplanted by a system of provincial departments of education in the expectation that these would not only produce a unified nonracial system but also improve the character of the system as a whole and lend itself to the "national effort."

The last thirteen years have seen a steady stream of reforms and a host of policies and strategies intended to institutionalize an education system of high quality. Yet these gains were hardly adequate given the challenges facing the avowed project of social "transformation." In the first place, this project, heralded by the African National Congress (ANC) in government, was hedged in by the intractably severe structural inequality that has characterized South African society for centuries. Compounding the problem were the self-imposed limits of the ANC's agenda of "national democratic" revolution (Alexander, 2002). Moreover, the possibilities for systemic change in education were dependent on how the constraining conditions for reform arising from outside the education system impacted them. Such analysis required an understanding not only of the broader historically engendered social pathology but also of the limitations of the very process of political settlement preceding 1994. Without such analysis the possibility for targeted and purposeful *social* interventions needed for genuine transformation was, as we have come to realize, truncated.

Beyond the Promise

Over the last few years there has been an abundance of data pointing to the high—even rising—levels of social inequality, income poverty, and unemployment. According to Statistics South Africa (2002), a statutory body, 27 percent of adults are unemployed. This is a narrow definition of unemployment and this excludes those who have given up looking for work. Survival strategies such as employment in precarious and poorly paid work in the informal sector are not considered in these unemployment statistics. The addition of such categories would increase the unemployment figures to catastrophic levels. The quality of jobs is also declining as permanent, secure employment is replaced by precarious and vulnerable forms of intermittent employment at low pay and without benefits. Many prefer using the more accurate, expanded definition of unemployment in South Africa, which is 41 percent (COSATU, 2006). Unemployment is highly racialized, gendered, and unevenly distributed by region. Of the unemployed, over 70 percent are under the age of thirty-five (ibid.). Black women and those in rural areas fare worse in comparison to men and people in urban areas.

Equally disturbing statistics exist in respect of income poverty and many other human development indices including health, sanitation, and mortality. HIV/AIDS has ravaged many working class communities as pharmaceutical companies block the use of generic medicine, and the state drags its feet in rolling out antiretroviral drug treatment.

The pervasiveness of this reality inspired from about 1998 the need for the reinsertion of the voice of those who felt most marginalized by the inability of the new state to address its pressing needs. This included the reemergence of community-based initiatives to confront the problem of disempowerment. Examples of these attempts are described only briefly here as space and time do not permit us to come to grips with the details of what was done to conceptualize, plan, design, and conduct community-based enquiry and to critically examine its purposes and outcomes. The few examples we refer to constitute an approach to social enquiry based on the premises argued earlier. We regard these studies as formative and developmental rather than as exhaustive explorations. A recent publication about community-based struggles for social change speaks more fully about the many and accumulating struggles of researchers and social activists engaged in and with communities in issues beyond educational struggles (Ballard et al., 2006).

The initial impetus for this activity was provided by the Poverty and Inequality Hearings organized by the South African Non-Governmental Organization Coalition (SANGOCO). Between March 31 and June 19, 1998, over ten thousand people took to the streets, participated in public hearings, and made submissions about their experiences of continued poverty and inequality in postapartheid South Africa.

The hearings were organized thematically and held in all nine provinces of the country. They dealt with employment, education, housing, health, the environment, social security, and rural development. These hearings were supplemented by background papers compiled by NGOs and research organizations involved in the different fields. The research focused on the legacy of poverty and inequality in each sector and its impact on people's lives, the extent to which current practices and policies contributed to improve conditions and recommendations on the measures required to assist groups to access their socioeconomic rights. In addition to the verbal testimonies, the Education Theme co-coordinators received scores of written submissions from parents, teachers, school governing body members, early childhood development and adult education providers and learners, student and youth organizations, trade unions, NGOs, and church groups. These ranged from the carefully worded arguments of research organizations to the poignant testimonies of some of the most marginalized demographics such as child workers and prisoners.

The hearings provided concrete evidence that the inability to afford school fees and other costs such as uniforms, shoes, books, stationery, and transport was one of the major obstacles blocking access to education. In some cases, parents or even the pupils themselves discontinued schooling as costs imposed too heavy a burden on the family. The lack of electricity, desks, and adequate water and toilet facilities in schools were also referred to in a number of submissions. Overcrowded classrooms continued to be a standard feature in poor communities. Frustrated by unfulfilled promises, many poor communities, particularly women in these communities, scraped together their meager resources in order to provide rudimentary education facilities in their communities.

The hearings brought public attention to these issues nationally and exerted pressure on government officials and politicians to reexamine and reverse contested strategies such as the government's neoliberal macroeconomic strategy (GEAR-Growth, Employment and Redistribution Strategy) with its negative effects on the promise of providing for basic needs and services. Held in the absence of grassroots community organizations to take forward the demands of the people, the hearings had only a limited impact. In fact, even three years after the Hearings, the Department of Education's School Register of Needs Survey (Department of Education, 2001), which quantifies the provision of physical infrastructure for South Africa's schools, continued to show that adverse conditions persist and in some cases have increased.[1]

A consequence of the macroeconomic strategy of the government, in which fiscal austerity was the key, saw the emergence of privatization of public services in a range of areas including education, health, housing, and even the supply of water. This was simultaneously a concrete expression of the increasing grip of human capital theory conceptions of development applied to the public schooling system and the extension of these through stringent budgetary constraints and fiscal austerity allied to marketization, "public-private partnerships," cost recovery, and cuts to education and other social services. These strategies were pursued simultaneously with rhetorical support for redistribution and redress (Samson and Vally, 1996), and expressed the contradictory impact of the politically "negotiated settlement" between the ANC (and its allies), the nationalist apartheid regime, and, most importantly, national and multinational corporate interests. This denouement ensured the continued and extended interests of old and new national elites, albeit through new configurations of power sharing with each other. Alexander (2006) writes:

> Ardent as well as "reluctant" racists of yesteryear have all become convinced "non-racialists" bound to all South Africans under the "united colours of capitalism" in an egregious atmosphere of Rainbow nationalism. The same class of people, often the very same individuals who funded Verwoerd, Vorster and Botha, are funding the present regime. The latter has facilitated the expansion of South African capital into the African hinterland in ways the likes of Cecil John Rhodes or Ernest Oppenheimer could only dream...In this connection, it is pertinent to point out that the strategy of Black Economic Empowerment—broad-based or narrow is immaterial—is no more than smoke and mirrors, political theatre on the stage of the national economy. The only way that erstwhile Marxist revolutionaries in the liberation movement can justify their support and even enthusiastic promotion of these developments is by chanting the no longer convincing mantra: *There is no alternative!* Hence, we need to examine this particular mystification and abdication of intellectual responsibility.

Over the past few years a new layer of cadre are insistent that there must be alternatives to neoliberalism and resistance against the status quo in South Africa is growing. New, independent grassroots social movements have formed and are establishing continuity with past movements. They are beginning to expose the hollowness of electoral promises around social delivery and corruption. They have also taken the lead in resisting neoliberalism in all spheres of life.

Subsequent to the poverty and inequality hearings of 1998, the staff from the Education Policy Unit together with activists from various social movements formed the Education Rights Project (ERP) in 2002. An important difference with the period of the Poverty and Inequality Hearings was the presence of nascent but increasingly expanding social movements. The ERP worked closely with these movements in its five campaign areas, namely, the cost of education, infrastructure and facilities, sexual harassment and violence, farm schooling, and adult basic education.

Like the earlier People's Education Movement (see Motala and Vally, 2002), the ERP's participatory research initiatives with the various emerging social movements and community organizations is a form of social accountability. It asserted the need for civil society to have access to collective self-knowledge, independent of government, in order to hold the state to account for its policies. It is used as a social check on the state's "numbers" and "statistics," which are forwarded by state functionaries as "official justification" for its policies, and, in this instance, the right to education. This critical research according to Kincheloe and McLaren (1998) "becomes a transformative endeavor unembarrassed by the label 'political' and unafraid to consummate a relationship with an emancipatory consciousness" (p. 264). Those in the ERP initiative see their research as, "the first step towards forms of political action that can address the injustices found in the field site or constructed in the very act of research itself" (ibid.).

The troubles and struggles of individuals and communities to educate their young in very trying conditions, to make the hard-won right to education in South Africa's constitution a reality, are vividly portrayed in these testimonies. In addition to collecting personal testimonies, a number of communities across the country have worked with the ERP to design tools and collect local quantitative data about the cost of schooling (e.g., school fees, books, school uniforms), violations of education rights, and basic household and community data.

The importance of such a research process is that it promotes democratic and cooperative practices in the production and the designation of what constitutes knowledge; it demystifies the research and facilitates a social and active response to complex policy issues. The outcomes of the research inform the design of a campaign aimed at improving local education. This will ultimately contribute to democratizing the debate on, for instance, the impact of government budgets on local education, as communities themselves will have the data to challenge or support assertions made by the state or other organizations about provisioning for education.

The ERP deliberately chose to structure its research through a process of direct collaboration and work with community-based organizations in the areas in which the research was to be conducted. Guided by this principle, initial collaborative work was carried out in two sites in the province of Gauteng. The sites were Durban Roodepoort Deep (DRD) and Rondebult. Both communities have relatively strong social movements affiliated to the Anti-Privatisation Forum (APF). The research endeavor in these communities was used as a model in subsequent work with communities throughout the country.

The research was conducted over the period 2003–2005.[2] In brief, the research consisted of house-to house surveys in the case of one site, DRD, where local youth activists conducted in-depth interviews with community members. In Rondebult, interviews, focus groups, and a short questionnaire were used to gather data. Members of the community were engaged throughout the development of the research process, its methodology and design and its implementation. With the assistance of the ERP, the data (in both these case studies and with other communities) were analyzed and then presented at mass community meetings where a discussion was held regarding actions to be taken to deal with the problems identified. This led, in the first instance, to the realization that democratic discussion and debate was critical to influencing social policy and practice and the choices made by government. It also reinforced the view that communities themselves required the data necessary to challenge or support assertions made by the state or other organizations about provisioning for education. In the case of DRD, this led to important concessions by the state regarding the costs of scholar transport. Moreover, it was realized that until communities are able to act autonomously, powerful interests opposed to the interest of working class communities will continue to be dominant in the making of social choices and their underlying values.

Durban Roodepoort Deep, also known as Sol Plaatjie, is a community living in what previously was a hostel or compound[3] for mine workers of the DRD Mines. The hostel is approximately a kilometer from a relatively new settlement known as Bramfisherville in Soweto. These residents had come to DRD in about January 2002, when almost twenty-five hundred families were evicted by the Johannesburg Metropolitan Council from an informal settlement known as Mandelaville in Soweto. Mandelaville has its genesis in the Soweto Uprising of 1976. The site on which the original Mandelaville settlement was built, long before the eruption of shacks, was a municipal building razed to the ground by students in 1976. From about 1977, a handful of homeless families and individuals began occupying the razed building. By the time of the forced eviction in 2002, some people had lived in Mandelaville for twenty-five years.

The residents had moved from Mandelaville on the assurance that DRD would be a temporary stop on the way to more permanent and properly serviced arrangements, including housing through the government's Reconstruction and Development Programme (directed especially at poorer communities). In the first few weeks of the settlement, nearly as many people were moving out as were moving into the settlement owing to dire conditions such as the absence of shops providing basic amenities, schools, clinics, and transport, and the presence of rampant violent crime. Approximately eight hundred families stayed, in the hope that while they waited for the promised houses, school transport would be provided and a temporary clinic would be built. From the preliminary data gathered through a door-to-door census of 763 households in the area, this community, fairly typical of similar communities around the country, evinced some clear characteristics.

- The average household income was R894 ($1 equaled 7 South African Rand at the time of going to press) per month. Fifty-five percent of households reported to have at least one member of the household in formal employment. Only 5 percent of households reported to having at least two people in formal employment. Twenty-one percent of the sampled households reported to having a member that was employed on part-time basis. This category comprised almost exclusively of part-time and casual workers—domestic workers and gardeners in and around the nearby suburbs. One hundred and twenty-six households reported that apart from income earned through employment, they did not receive any other income.
- Just less than 50 percent of households reported to be paying for child school transport. These households paid an average of R115 per month for scholar transport. A hundred and thirty-one households reported to have a second child at school for whom they also needed to pay scholar transport. The mean for transport fares for the second child was very close to that of the first child—R118 for the first child compared to R116 for the second. In other words, more than 50 percent of households paid up to R230 per month in transport fares.
- The average school fees in DRD were R201 per month. There is large variation in the amount of school fees that households pay reflecting the number of children of school-going age in each household, variation in school fee charges in different schools, and the inability of some households to pay school fees. Seventy-one households had at least two children at school and paid an additional average of R87 per month in school fees for the second child. Households with at least two children at school paid an average of R288 on school fees per month. The foremost cited reasons for not attending school were the cost of school fees and transport.

In February 2004, the Education Rights Project, the Anti-Privatisation Forum, and an Education sub-committee formed by the community of DRD cohosted a workshop for caregivers with children of school-going age. The workshop sought to stimulate discussion about the right to basic education, to gain insight into the community's view of the relation between education and imaginations for community development, to identify key barriers to the right to basic education, and to develop plans around local action. Most importantly, the workshop was also used to consider the methodology for community participation and the nature of the data required to investigate issues relating to the right to basic education.

The group identified several barriers to basic education. First, the existence of undemocratic and unrepresentative school governing bodies was noted by many. It was reported that members of a school board of governors tended to be drawn from caregivers who could afford to pay school fees, marginalizing the views and interests of very poor parents. Second, the cost of schooling was seen as particularly onerous. For this group the burdensome costs were, in the following order, scholar transport, school uniforms, and school fees. Third, the use of corporal punishment in schools was pervasive. Despite the official prohibition against corporal punishment, the participants believed that this practice was still common, especially in primary schools. Finally, internal community migration and forced removals exerted a negative effect on schooling. Many participants reported that their children had lost school years as a result of poverty induced internal migration from city to city and of forced municipal removal of shack dwellers by the state.

Community members also reflected on the conditions that could improve the quality of schooling. Their proposals included the construction of a school within a reasonable walking distance from where children lived, state provision of at least one meal a day for school children, and free stationery and textbooks for all grade levels. Suggestions about how to confront and deal with these barriers included the need for all adults in the community to inform themselves of the education rights enshrined in the constitution and other legislation. A call was made for a public meeting with local education officials where the community could raise its concerns, suggestions, and demands. The media was identified as a potentially useful partner both showing the poor conditions under which people lived and as a means of increasing pressure on the government to act. The group emphasized the importance of independent self- organization, mobilization, and public protest by the community itself. In their reflections in the middle of 2005, after much struggle and protest, members of the community said:

> We do not have free and quality education in this community. Whatever gains we have made for our children around school transport and the erection of the primary school is what we have fought for ourselves.[4]

> We are concerned about the safety of our children. I cannot say that we are happy about the school that has been recently erected near DRD because the fact is our children have to walk some distance to this school. They [i.e., government] must not pretend that they brought this school because they wanted to; they are responding to the pressure of our children. The reason why I do not approve of the school being outside this camp is that this is not a safe place, and children can get murdered or raped on their way to school. Even more important, if this school was inside this camp we could meet and talk to the people that are teaching our children—at the moment they are strangers to us. I am an old man now and I do not have any children at school. But I am still a concerned citizen, and it pains me to see children not at school. Because here, in this camp, we are expected to produce doctors, lawyers, ministers...you name any top position...presidents, who can understand the situation of the poor.[5]
>
> I really do not think that the right to education applies to my community—because children that need to be at school are not at school because they cannot pay school fees. It really does not matter that there is a law about school fees exemptions because schools are unwilling to implement it.[6]
>
> I am not sure why some children do not go to school. It may be because the parents do not have money for food and therefore cannot concentrate in class because they are hungry. With regard to adults I know that many of them, just like me, would like to go to school but there is no opportunity, however, to do so in this community.[7]
>
> I think that a reason for why some children are not at school has to do with the eviction from Diepkloof. People lost birth certificates and report cards. These are difficult to get back when you do not have money to make endless trips to the Department of Home Affairs.[8]

These views of members of the community should be contrasted to the "deficit" view of a state official who typically responds in a way many community activists have labeled as callous, aloof, and indifferent:

> I do not know why children are not at school. I think you should ask their parents; they might be in a position to answer your question. There are buses and there is a school nearby—what more do you want from government? If they do not want to take a bath and go to school, how am I supposed to answer that question? I do not know whether or not buses will be provided for the children next year; if they are not provided I would have little influence on that issue because it does not fall under my department.[9]

Rondebult was established in 1998 as part of the government's Reconstruction and Development Programme (RDP). It is about five kilometers from Leondale/Spruitview to the east of Johannesburg. The people were resettled from an old and well-established township with schools, clinics, and shopping centers. Similar to other resettled communities, the people of Rondebult were adversely affected by the effects of the relocation. These include the impact on their livelihoods and the breakup of social networks and family relations built over many years, as well as the absence of any health care services and schools in the area. Children have to travel long distances to access schooling. The majority of people in the community are unemployed. A large number of those who are employed work as domestic servants and casual workers in nearby suburbs and industrial areas.

As in DRD there were no schools, and the only available options for the community were to allow their children to walk the long and dangerous distances to school, pay the expensive scholar transport, or keep the children out of school. Here too, engagements between the ERP and the community soon revealed that barriers to education in the area were not just limited to the issue of transport. The other factors that militated against the right to basic education were school fees, the cost of textbooks and stationery, and child hunger.

In a random sample of caregivers, 75 percent reported to be out of work. Only in a single instance were both caregivers in a household reported to be in full-time employment, and even in this case they could afford neither the school fees nor the school bus fares. For the rest of the caregivers who reported to have had some kind of productive employment, most reported that they were involved in informal trading or worked on a part-time or casual basis for anything ranging from one to three times a week. Informal trading and domestic service work generate very little income, typically between R50 and R100 per day.

Of the respondents, 97 percent paid R400 per year on school fees. Transport fares were reported to be R40 per month for each child. Average school uniforms costs per child was R200 per year. The average household had 2.5 school-going age children, translating to average spending on school uniforms of R500 per year. Added to this were the costs of textbooks, depending on the school and grade level, of R90–150 per annum. Based on this data, an average Rondebult household spent a massive 33 percent of its total earnings on schooling.

These case studies have assisted the ERP and social movements to pressure the state into reviewing the formula for fee exemptions and have strengthened the capacity of communities to organize around violations of education rights. As a consequence of massive campaigning and lobbying by social movements[10] and an intense press campaign against the school fees policy, in September 2002, a review of the financing, resourcing, and the costs of education was announced. The government set up a reference group of twenty-seven members, consisting of a core team from the DoE and "prominent economists and managers from inside and outside government" (DoE, 2003, p. 8), as well as from the World Bank.

Although the ostensible purpose of the review was to "stimulate and inform constructive discussion" on how government schools are resourced, the Review Report was formulated amid numerous complaints by labor- and community-based organizations, who charged that there was no participation by any representatives of education unions, school educators, governing bodies, and community organizations (APF submission, April 2003). In addition to the lack of participation by key groups, critics argued that the Review was not adequately publicized to encourage wide-ranging responses and that the time frames for submissions did not allow for democratic processes to run their course.

Education, Poverty, and Social Marginalization

These enquiries provide evidentiary proof about the effects of the policy choices of government—especially through its orientation to the costs of education. In addition these enquiries also indicate that school fees constitute a real barrier to the right to education for working poor, in which women in particular bear the greatest burden of such oppressive and discriminatory practices. Until very recently, government remained steadfast on a fee policy that inevitably led to the exclusion of many children from schools. The reason for this seems to be the unbending resolve of policy makers to pursue conservative macroeconomic policies in which the choice of austerity measures outweigh the imprimatur of the rights enshrined in the Constitution. These measures, whose inspiration lies largely in the monetarist discourses, policy advice, and practices of agencies such as the World Bank, trump all other considerations regardless of history and context.

The implication of this is that the right to education will remain largely unrealized for the rural poor and working class children and systemic inequalities and disparities will continue to exist for both girls and boys not only because of factors endogenous to the educational system itself but also because of the larger and external constraints on the exercise of the right to education. Despite the greater educational access attained in South Africa, compared to most of the

African continent, education inequity continues to be a fundamental and systemic characteristic of the system.

These case studies are consistent with the results of other studies and confirm the reality that poverty and poor public services' provision in poor communities are inextricably linked. It implies also that reforms that are directed at the educational system alone are not adequate. The structural character of poverty and inequality cannot be resolved alone through education policy reform including the equalization of educational finance across the system because of the externalities that constrain the conditions for educational access and, most importantly, educational quality.

Educational interventions remain important but are partial in relation to the social outcomes of education and the goal of social transformation—that is, the transformation of South African society from among the most unequal societies on earth to a more equal society. What is required is a broader and more purposeful approach to social reform and redistributive strategies, a clearer orientation to the underlying values of society, to issues of wealth and income distribution and social empowerment, and the relationship between a purposeful state and civil society. Only this is likely to deal with the deeply entrenched social pathologies that affect South Africa and other developing societies and secure the gains made through social policy in developed nations.

Engaged Research and Community Activism

More important perhaps is the suggestion implied in these studies that a large part of the problem confronting such communities is the inability of decision-makers to construct a proper dialogue about the complex relationship between the delivery of services and the persistence of poverty. The approaches adopted in these studies speak to their public purposes in relation to communities in which poverty and oppressive conditions are pervasive, the second approach relating to how social reform policies are conceptualized. They highlight the importance of community participation through independent social movements in campaigning for access to quality education and other social goods. They constitute a critique of more conventional approaches to reality, in which the certainties of academic approaches appear to be taken uncritically, and debilitate the communities of the poor and oppressed.

In criticising the frequent dissonance between research and action, Jean Dreze (2002), a long-time collaborator with Amartya Sen on works dealing with public action by community groups in India, has this to say:

> Social scientists are chiefly engaged in arguing with each other about issues and theories that often bear little relation to the world . . . The proliferation of fanciful theories and artificial controversies in academia arises partly from the fact that social scientists thrive on this confusion (nothing like an esoteric thesis to keep them busy and set them apart from lesser mortals) . . . To illustrate, an article in defence of rationality (vis-à-vis, say, postmodern critiques) would fit well in a distinguished academic journal, but it is of little use to people for whom rational thinking is a self-evident necessity—indeed a matter of survival . . . It is no wonder that "academic" has become a bit of a synonym for "irrelevant" (as in "this point is purely academic"). (P. 20)

Dreze is at pains to show that he is not dismissing the importance of academic rigor but that scientific pursuit can be enhanced even further if it is grounded in "real-world involvement and action" (p. 21).

Academic research is largely disdainful of the importance of the voices of marginalized communities, and their rich and valuable experience is equated with "anti-science," its traditions, history, and culture effaced from public view, and all of it regarded as irrelevant to policy analysis.

Removing the experience of those most affected by the direct consequences of the dominant "solutions" to the problems of developing societies means a reliance on conceptions based largely on analyses that have no orientation to the relevance of their histories, culture, tradition, and value systems. Rather, these analyses rely in the main on contested bodies of data, poor conceptions of issues such as poverty and inequality,[11] a-contextual theorizations and political and economic agendas to manipulate the outcomes of development in the interests of particular ideological interests. In the social sciences, most of these analyses continue to rely, in one form or the other, on the discourses of modernization that grew out of the ideas of Rostow (1960) and Parsons (1966)although quite clearly they represent advanced variations on these ideas. They rely on models based on inappropriate contexts and can do so because of the imperative nature of their relationship with dependent states.[12] This is so despite the ostensibly "empowering" intentions of such interventions. As Mamdani (1996) argues:

> A unilinear social science . . . tends to caricature experiences that are "residual" to its standpoint while at the same time it "mythologizes"

> its own experiences. If the former is rendered ahistorical, the latter is ascribed a suprahistorical trajectory of development, a necessary path whose main line of development is unaffected by struggles that happened along the way. (Pp. 9–10)

A critique of such analyses, paraded as "innovative knowledge," raises questions about the power and relevance of particular knowledges based on such conceptualizations and about their effects on the countries of the "South" and the conditions for their democratic development. It also raises questions about the need for reexamining such knowledges with a view to postulating alternative approaches to the development of ideas through an examination of the deeper and directly relevant sources of knowledge. Such a deeper and socially informed approach would have to burrow beyond the epistemological dependencies written into South African history and search for new modes of thinking respecting "other" and contextually relevant viewpoints and concepts.

In discussions about the context of changing societies in the developing world very little attention, if any, is paid to the importance of the experiences and knowledge that lie deep in the reservoirs of the lived reality and reflection of these societies. Even less attention is paid to the constitutive struggles they undertake. Yet the history of ignoring such knowledges is replete with the skeletons of failed "development" and other projects based on the conceptions, theories, and practices that are totally inappropriate for the possibilities of genuine development. The failed discourses of modernization continue, despite this reality, to be purveyed as salient to the conditions prevalent in structurally underdeveloped social formations, and strong global institutions continue to put huge resources into these failed modes of development.

Our intention here is neither to romanticize the value of local, traditionally based, indigenous, or other epistemologies nor to argue the case for a postmodern discourse. Nor are we attempting to argue from a "communitarian" viewpoint that Mamdani defines as an "Africanist" perspective based on the "defense of culture" in which the "solution is to put Africa's age old communities at the center of African politics"—communities that have been "marginalized from public life as so many 'tribes'" (p. 3). We are also mindful of and agree with Hobsbawm's (2004) view that:

> The major immediate political danger to historiography today is "anti-universalism" or "my truth is as valid as yours, whatever the evidence."

> This naturally appeals to various forms of identity group history, for which the central issue of history is not what happened, but how it concerns the members of a particular group. What is important to this kind of history is not rational explanation but "meaning," not what happened but what members of a collective group defining itself as outsiders—religious, ethnic, national, by gender, lifestyle or in some other way—feel about it. This is the appeal of relativism to identity-group history. (P. 5)

We agree also with his argument that against this "endless claptrap and further trivia" of "in-group histories" it was as necessary to reassert a belief in history as necessary to an enquiry about "the course of human transformations" and to deal with the problems of distortion "for political purposes" as it was to deal with the claims of the "relativists and postmodernists who deny this possibility" (ibid.).

We would add however that for such reasoned enquiry about "what happened" to take place, it was necessary, in the first place, to seek a restitution and recognition for the heritages, normative value systems, legacies, and experiential knowledges of communities that have hitherto been ignored in the construction of their destinies. It was, as Amilcar Cabral (1973) said, necessary to "return to the source" even while we acknowledge that such a source too must invariably contain the paradoxes of local and regional histories, the markings of its own internal struggles and of its engagement with other knowledges, both the richness and the poverty of some of its expositions, the claims of social hierarchies, and the diversity of its voices. Yet this experience and knowledge too have an insistent right and a claim to relevance and legitimacy because of their directness and testable authenticity.

Rarely is it understood that these voices too could make a real contribution to the discourse of democratic development, to finding solutions, through engaging in participatory research about the nature of the issues confronted by communities. The role of these voices as potential agents for social change is wholly ignored, and this has consequences for how research is conceptualized and practiced.

Yet a number of new social movements have used the democratic space available today to increasingly create a groundswell of support for resistance around education. The praxis of these organizations is based on an understanding of democratic citizenship that speaks to peoples' lived experiences. Increasingly silent apathy and hopeless resignation is giving way to creative initiative and courageous attempts by young people and their parents to continue the long South African traditions of democratic participation from below, now supported by research that has a dialogical relationship with this resistance.

Notes

1. Of the 27,000 schools in South Africa the study estimated that 27 percent had no running water, 43 percent were without electricity, 80 percent were without libraries, and 78 percent had no computers. A total of 12,300 schools used pit latrines and 2,500 had no toilets at all. In schools that did have toilets 15.5 percent were not in working order. Schools requiring additional classrooms numbered over 10,700. According to the survey the number of state-paid educators decreased dramatically by 23,642 while school governing body-paid educators, almost exclusively in the schools for middle and upper class communities, increased by 19,000 signaling a decline in state funding while creating a labor market in the supply and distribution of teachers.
2. The details of the methodological approaches and research design, a detailed report of the data collected and its limitations are not dealt with in this chapter and are available in the report done for the purpose (see Vally and Ramadiro, 2007).
3. These constituted dormitory bed-spaces for migratory male workers under the apartheid system and ensured direct controls over the lives of these workers.
4. Percy Khoza, community leader and school bus organizer October 12, 2005 (translated from isiZulu).
5. Tata Rabbi, elderly community and religious leader, September 12, 2005 (partially translated from isiXhosa and isiZulu).
6. Daniel Dabula, deputy chairperson Mandelaville Crisis Committee, July 11, 2005 (translated from isiZulu).
7. Dudu Dube, leader of a women's forum, September 12, 2005 (translated from isiZulu).
8. Percy Khoza, community leader and school bus organizer October 12, 2005 (translated from isiZulu).
9. Councillor Paulos Mahlabi, ward 17, July 11, 2005.
10. The demands for a review of funding came from a variety of civic and social organizations—chief among them were civil society groups, student and community organizations who were key in boycotting school fees; the Anti-Privatisation Forum; the Anti-Eviction Campaign; the Global Campaign for Education; the Education Rights Project; and the South African Democratic Teachers Union. It is worth underscoring the role that social movements played in pressurizing the government for a comprehensive review, particularly because their absence from the Review Committee and its deliberations is striking.

11. See du Toit (2005, p. 1) wherein it is argued that in relation to the measurement of poverty "econometric approaches to chronic poverty are dependent upon mystifying narratives about the nature of poverty…they direct attention away from the underlying structural dimensions of persistent poverty."
12. See Ohiorhenuan (2002) in this regard, particularly to the role of the World Bank and IMF.

References

Alexander, N. (2002). *An Ordinary Country.* Pietermaritzburg: University of KwaZulu-Natal Press.

———. (2006). South Africa today. The moral responsibility of intellectuals. Lecture delivered at the tenth anniversary celebration of the Foundation for Human Rights in Pretoria, November 29, 2006.

Anti-Privatisation Forum (APF). (2003). Response submission to ministerial review of financing, resourcing and the costs of education. Unpublished submission.

Appadurai, A. (2006). The right to research. *Globalization, Societies, and Education,* 4(2).

Arjun Appadurai, The right to research, *Globalisation, Societies and Education*, 4(2), 167–177.

Ballard, R., Habib, A., and Valodia, I. (Eds.). (2006). *Voices of Protest: Social Movements and Post-Apartheid South Africa.* Pietermaritzburg: University of KwaZulu-Natal Press.

Bhorat, H., and Kanbur, R. (2006). Introduction: poverty and well-being in post-apartheid South Africa. In H. Bhorat and R. Kanbur (eds.). *Poverty and Policy in Post-Apartheid South Africa* (pp. 1–18). Cape Town: HSRC Press.

Bond, P. (2006). Lack of service delivery is turning poor settlements into hotbeds of activism. *Sunday Independent*, January 29.

Cabral, A. (1973). *Return to the Source: Selected Speeches by Amilcar Cabral.* New York: Monthly Review Press.

Cammarota, J., and Fine, M. (2008). *Revolutionizing Education.* New York: Routledge.

Cooper, L., Andrew, S., Grossman, J., & Vally, S. (2002). 'Schools of labour' and 'labour's schools': Worker education under apartheid. In P. Kallaway (ed.). *The History of Education under Apartheid, 1948–1994: The Doors of Learning and Culture Shall Be Opened* (pp. 111–133). Cape Town: Maskew Miller Longman.

COSATU. (2006). Possibilities for fundamental social change. Political Discussion Document, unpublished. Online at http://www.sacp.org.za/main.php?include=docs/misc/2006/cosatu_pol_disc.html.

Department of Education (DoE). (1996). *South African Schools Act.* Pretoria: Government Printers.

Department of Education (DoE). (2001). *Report on the School Register of Needs Survey.* Pretoria: Government Printers.

Department of Education (DoE). (2003). *Review of Financing, Resourcing and Costs of Education in Public Schools.* Pretoria: Government Printers.

Dreze, J. (2002). On research and action. *Economic and Political Weekly*, 9, xxxvii.

Du Toit, A. (2005). Poverty measurement blues: some reflections on the space for understanding "chronic" and "structural" poverty in south Africa, paper prepared for the First International Congress on Qualitative Enquiry, University of Illinois at Urban Champaign, May 5–7. Paper ID 1–295.

Hobsbawm, E. (2004). *Asking the Big Why Questions.* Speech to the British Academy colloquium on Marxist Historiography, December, London, UK.

Kincheloe, J.L., and McLaren, P. (1998). Rethinking critical theory and qualitative research. In N. Denzin and S. Lincoln (eds.). *The Landscape of Qualitative Research Theories and Issues* (pp. 138–154). Thousand Oaks: Sage.

Mamdani, M. (1996). *Citizen and Subject.* Princeton, NJ: Princeton University Press.

Motala, E. (2007). Engaged social policy research: reflections on the nature of its scholarship. Unpublished paper. Museums and Universal Heritage. Universities in Transition—Responsibilities for Heritage UMAC's 7th International Conference August 19–24, 2007, Vienna, Austria, within the ICOM General Conference General information: http://www.icom-oesterreich.at/2007/index.html

Motala, S., and Vally, S. (2002). From people's education to Tirisano. In P. Kallaway (ed.). *The History of Education Under Apartheid* (Chapter 5, pp. 174–194). London: Peter Lang.

National Assembly. (2007). *Internal question paper* (No. 43/2007), November 22.

Ohiorhenuan, J. (2002). *The Poverty of Development; Prolegomenon to a Critique of Development Policy in Africa*, Sixth Professor Ojetunji Aboyade Memorial Lecture, September 25.

Parsons, T. (1966). *Societies: Evolutionary and Comparative Perspectives.* NJ: Prentice-Hall Englewood Cliff.

Rostow, W. (1960). *The Stages of Growth: A Non-Communist Manifesto.* Cambridge: Cambridge University Press.

Samson, M., and Vally, S. (1996). Snakes and ladders: the promise and potential pitfalls of the national qualification framework. *South African Labour Bulletin* 20 (4), 7–53.

Statistics South Africa. (2002). Income and expenditure of households in South Africa. *Statistical Release Poll.* Pretoria: Government Printers.

United Nations Development Programme (UNDP). (2004). *South African Human Development Report. The Challenge of Sustainable Development.*

Oxford: Oxford University Press. Online at http://www.undp.org/eo/

Vally, S., and Ramadiro, B. (2007). *The Social Movements and the Right to Education in South Africa: Selected Case Studies from the Education Rights Project (ERP)*. Centre for Civil Society University of KwaZulu-Natal. KwaZulu-Natal: University of KwaZulu-Natal Press.

Chapter 3

Taking on the Corporatization of Public Education: What Teacher Education Can Do

Pepi Leistyna

Most citizens of the United States are unaware of the extent to which public schools are controlled by private interests such as publishing, food, and pharmaceutical companies, for-profit education management organizations, and corporate lobbyists (Kohn and Shannon, 2002; Molnar, 2005; Molnar et al., 2006; Saltman, 2000, 2005). The stealth onslaught of privatization and commercialization of this vital institution should come as no surprise given that education reform over the past few decades has been masterminded, in large part, behind closed doors, by a handful of corporate executives, politicians, and media moguls who have already profited handsomely from the over six hundred billion dollars a year education industrial complex (Bacon, 2000; Gluckman, 2002; Humes, 2003; Miner, 2005; Olson, 2005; Rees, 2000).

It's a pretty simple equation: when you have a captive audience, given that K-12 education is mandatory in the United States, private interests within the logic of capital can't help but salivate and pounce. But as all good capitalists know, the overriding objective of corporations is to maximize their profits. So what they have to do in order to shape public policy in their own interests—and gain consent on those rare occasions when the general public is involved in the process—is disguise their "profit over people" mentality by wrapping themselves in an image of expertise and compassionate concern for the education and future of our children.

To illustrate this point, the following research takes a look at standardized high-stakes testing in the wake of the No Child Left Behind (NCLB) movement. Although it is well beyond the scope of this chapter, it is important for concerned citizens to look into the various problems that high-stakes testing engenders; for example,

- most school administrators, parents, and communities are stripped of any substantive influence over the educational process;
- educators are disempowered by what publishing companies describe as "teacher proof" materials;
- in order to teach to the test, many schools have been forced to drastically narrow their curriculum;
- under extreme pressure to produce results or face losing their jobs, cuts in federal resources, and school closure, educators often lower standards and engage in unethical behavior in order to raise test scores;
- research clearly shows that racism and the structures of social class affect the quality of faculty and staff employed across school districts, the overall allocation of resources to schools, as well as the content of assessment instruments and testing results;
- nationwide, there are drastic retention and "drop out"/push out rates that are the direct result of this practice;
- there is a large body of research that unequivocally shows that standardized tests should not be used to make critical decisions such as eligibility for graduation (Bacon, 2001; CNN Presents, 2005; FairTest, 2006; Gluckman, 2002; Haney, 2000; Kohn, 2000; Sacks, 2001; Sturrock, 2006; TC Reports, 2000).

While these are all extremely important issues, the general focus of this chapter is on the political economy of the testing industry; that is, a look into the ownership, intent, and regulation of the private forces that produce, provide materials, prep sessions, and tutorials for, evaluate, report on, and profit from these tests. By looking at and analyzing this powerful sector and the old-boy networks therein it is much easier to understand the synergy that exists among government, corporations, and the media and how this force has been controlling public education in the United States to the detriment of its people. It is from this position of awareness that activists can better mobilize against these forces, and teacher education programs can be restructured, if they are really intent on democratizing schools, in order to better meet the needs of educators and students alike.

A Bit of History

While standardized assessment has been around for a long time in the United States, its focus on student performance and accountability and its impact on grade promotion, eligibility for graduation, and overall school evaluation has increased dramatically. Over the past fifty years, expenditures on testing have increased by 3,000 percent (Frontline, 2006).

Many of the roots of the current assessment craze can be traced back to two significant events: the 1957 launch of the Sputnik satellite in the former Soviet Union—an achievement that was used to call into question the intellectual capital of youth in the United States; and the 1966 release of The Coleman Report—Equality of Educational Opportunity—which initiated a major shift in national strategy from targeting resources and their impact on student achievement to a focus on measuring individual performance.

Assigned the mission of mining data in an attempt to document the educational achievement of students in public schools, the National Assessment of Educational Progress (NAEP) was established by the federal government in 1969. Also known as "the Nation's Report Card," NAEP continues to be the only federal body that assesses student progress in all of the major subject areas.

While the Minimum Competency Test would appear on the scene in 1975, what really fueled the drive to increase federal and state funding for the standardization movement, and its privatization, was the Reagan administration's Commission on Excellence in Education. Intent on privatizing all public institutions and resources, and swamped with domestic and foreign policy problems, conservatives were dead set on distracting the public away from their disastrous programs and practices through a strategic focus on what they presented as the failure of public education in the United States. Using the old Cold War tactic of fear, it released A Nation at Risk in 1983. While the sources that informed the report's conclusions were vague to say the least, in no uncertain terms its authors drew a bleak picture of a country replete with semi-literate youth who were rendering the United States vulnerable in a competitive and ever-growing global economy and in a world "still threatened" by what they readily evoked as the "evil empire."

In bed with powerful members of the profit-minded business community, the Reagan administration called for a dramatic increase in the use of standardized curricula and assessment nationwide. It did so while simultaneously working to disarm teacher unions (all organized labor for that matter) and dismantle the Department of Education. The obvious goal was to have the private sector take the reigns of this reform movement as well as the lion's share of the federal and state monies directed toward public education. Their efforts proved successful in opening schoolhouse doors to corporate-sponsored education organizations that would carve a path for standards, measured achievement, and accountability that would lead the way to No Child Left Behind.

Groups like the National Center on Education and the Economy (NCEE) and America's Choice would emerge and campaign around the country offering their reform advice. Many of these organizations would charge school districts huge sums of money for membership fees and curricular and assessment materials. By the 1990s, they would also be active in forging lucrative partnerships with major publishing companies.[1]

Throughout the 1980s and 1990s, big business representatives and government officials would meet on a regular basis and strategize ways to forge coalitions and officially infuse standards nationwide. Organizations like the Business Roundtable, the National Alliance of Business, the U.S. Chamber of Commerce, the National Association of Manufacturers, and the American Business Conference would join forces and lobby heavily for this cause (Committee on Education and the Workforce, 1999; Mickelson, 2006).

After the National Education Summit in 1996 in Palisades, New York, a resource center "for states to help benchmark state academic standards and assessment against the best national and international exemplars" was established by a newcomer on the block, Achieve, Inc., led by governors and business leaders from around the country (Edward B. Rust, Jr., as cited in Committee on Education and the Workforce, 1999).

Perhaps the most blatant conflict of interest that surfaced in the early stages of this movement, foreshadowing things to come, was the development and influence of the New American Schools Development Corporation. Initially funded by IBM CEO Louis Gerstner, who threw in a cool one million dollars in order to get the organization on its feet, New American Schools was run by then president George H.W. Bush's deputy secretary of education, David Kearns—the former CEO of Xerox Corporation (Suchak, 2006). NASDC was successful in using its connections in Congress to federally fund its "reform package," and it also profited enormously from the purchase of its methodologies, curricula, texts, and testing materials by schools all over the country. As David Bacon (2000) notes, by the end of the millennium, forty-nine states were caught up in some way or another in the standards game.

Rather than look to seasoned educators for advice about their reform efforts, corporate-sponsored activists (who have always understood that a great many teachers and parents are diametrically opposed to their agenda) turned to the likes of the late Edward B. Rust, Jr., the chairman and CEO of State Farm Insurance. Regardless of the fact that he had no professional experience in education, and that his company was convicted of fraud and ordered to pay over $1.5 billion in damages in the late 1990s (France and Osterland, 1999), Rust nonetheless became the go-to-guy for spreading the word about the virtues of a market-driven educational system and the role that a corporate model of management should play therein. In fact, he would chair or direct all of the major committees, boards, and organizations of the private sector that supported testing, and he would testify in front of the federal government in favor of more rigorous standards (Committee on Education and the Workforce: U.S. House of Representatives, 1999). As Suchak (2006) notes, for his loyalty, he was not only elected to the Board of Directors of McGraw-Hill in 2001 (a point that will be of interest when I discuss the connections between McGraw-Hill and the Bush family), but "Almost simultaneously, Rust was chosen to head the incoming Bush Administration's Transition Advisory Team on Education."

No Child Left Behind Makes its Way to Center Stage

In his bid for the presidency in 2000, George W. Bush, as governor of Texas (1994–2000), effectively used what was referred to as the "Texas Miracle" to spearhead his educational policy plans based on high-stakes testing. Declaring himself the "education president," Bush would also use the "miracle" to help push through the No Child Left Behind legislation.

The Texas Education Agency (TEA) and Texas school districts had implemented standardized assessment to measure student knowledge of the state's curriculum in 1980. By 1990, the Texas Academic Assessment Skills (TAAS) program had been developed—through a five-year coordinated effort between TEA and the publishing giant Harcourt Brace—in order to measure students' abilities in writing, mathematics, reading, science, and social studies. The exam would be fine-tuned over the years to meet the Texas Essential Knowledge and Skills curriculum standards that were approved by the State Board of Education in 1997. Success on the TAAS is required in order to graduate from high school in Texas.

While spokespersons for conservative organizations, studies funded by advocates of the testing industry, and much of the mainstream media raved about the work being done in Texas, and the fantastic decrease in drop-out rates and increase in academic achievement that occurred since TAAS had been implemented, the so-called miracle was in fact a scam (CNN Presents, 2005; Gluckman, 2002; Haney, 2000; Humes, 2003; McNeil, 2000).[2] Those students who were perceived as potentially lowering the overall test scores were retained in grades where testing was not required, especially ninth grade; or they were placed in special education classrooms or labeled Limited English Proficient (LEP) and were thus exempted from taking the exam. In 1999 alone, forty-five thousand LEP students were not required to take the TAAS (Texas Education Agency, 2000). This language policy also exempts LEP students "if the child's language proficiency assessment committee determines that the child does not have sufficient language skills to take either the English or Spanish version of the TAAS" (Texas Education Agency, 2000). As any seasoned educator immediately recognizes, this leaves plenty of room open for inappropriate evaluation and exemption.

All of these aforementioned conditions, coupled with a long list of discriminatory practices that predate the standardization movement, but continue to fester, have caused high drop-out rates in Texas, and throughout the country for that matter.

It's not as if Rod Paige, the superintendent of the Houston Independent School District (HISD), could have been unaware of allegations of discrimination related to the TAAS exam or the mass exodus of students from his schools, given that he was elected to the district's Board of Education in 1990, had been the superintendent of HISD since 1994, was listed by *Inside Houston Magazine* as one of "Houston's 25 most powerful people in guiding the city's growth and prosperity,"[3] and the fact that in October 4, 1995, the Office for Civil Rights (OCR) received a complaint filed by the National Association for the Advancement of Colored People (NAACP) (Texas NAACP, 1995).

High drop-outs rates are fantastic for raising test scores but they simultaneously call into question the overall success of the standards program in place; so while they are often encouraged by corrupt administrators, they need to be disappeared from public view. In a self-serving malicious move, Paige and members of his administration cleverly manipulated the numbers and claimed that the drop-out rate of local schools was 1.5 percent, rather than the actual figure of over 40 percent (Sixty Minutes II, 2004). In reality, Texas schools, in particular those in Houston—the seventh largest school district in the country, have some of the worst drop-out rates nationwide.

There were also widespread stories of encouraging students to cheat on the TAAS in order to raise the scores (CNN Presents, 2005), as well as reports of the monopolization of the entire curriculum by prep classes that were teaching solely to the test (Haney, 2000).

Praising Paige for his "success," the publishing giant McGraw-Hill, which was making money hand over fist in Texas, awarded him its highest honor for educators, and in 2001 Paige was named National Superintendent of the Year by the American Association of School Administrators, and was appointed secretary of education by President George W. Bush.

Newly elected, Bush immediately, and with bipartisan support, signed into law the Elementary and Secondary Education Act of 2001, better known as No Child Left Behind. In a hitherto unheard of transfer of power to federal and state governments—granting them the rights to largely determine the goals and outcomes of these educational institutions—high-stakes testing was officially mandated nationwide.

NCLB required that by 2003, students in third through eighth grade be evaluated in mathematics and reading, and then reevaluated once in high school. By 2007–08, federal requirements also demanded that states administer tests in science in elementary, junior high, and high school. By the year 2014, all students must be proficient in these subject areas. Schools that don't meet this criteria will be stripped of their government funding, threatened with closure, or placed in the hands of charter schools or other such private management companies.

When you consider the fact that NCLB is already underfunded by more than forty billion dollars, a chilling reality sets in: all of this rhetoric about accountability, efficiency, effectiveness, and excellence in public education is really an ideological trap intended to ensure that public schools fail, thus paving the way for their complete privatization. Some states are well aware of the fact that these new mandates cost far more to meet than they could ever possibly financially provide for and have thus decided to opt out of their eligibility for federal funding and go it on their own.

Show Me the Money Trail

Embracing what is in fact an old neoliberal approach dressed up as innovative reform, the political machinery behind NCLB has effectively disguised the motivations of a profit-driven industry. Public schools now give nearly fifty million tests a year and the annual value of this market ranges from $400 million to $700 million (Frontline, 2006). The General Accounting Office estimates that by 2008, up to $5.3 billion will be spent by states trying to meet the requirements of this legislation (Miner, 2005). However, this figure doesn't include the enormous costs of prep sessions, practice tests, scoring and reporting, data storage, and let's not forget the nearly $7 billion a year market for instructional materials. But the enormous expense doesn't end there. Ben Clark (2004) reminds us of another lucrative market:

> Under NCLB, if a school fails to improve math and reading test scores within three years, a portion of its federal funding will be diverted to "parental choice" tutoring programs... These outsourced programs are run by private companies such as Educate Inc. owner of Sylvan Learning Centers whose revenues have grown from $180 to $250 million in the past three years and whose profits shot up 250% last year.

The potential for funneling taxpayers' money into private pockets is astounding. This is precisely why it is important for the public to watch closely how money is earmarked when politicians increase the federal budget for education: what often appears to be a concerned call to increase spending to improve schools for our nation's youth is actually a ploy to increase profit potential for those kingpins playing the standards game.

It is also important to know who the actual power brokers are that are lobbying government officials and consequently reaping the benefits of this movement. While many corporations (e.g., Edison Schools/Newton Learning Corporation, Educational Testing Service's ETS K-12, Advantage Learning Systems, Measured Progress, Data Recognition, Questar Educational Services, Kaplan, Princeton Review, BP, AT&T, Tribune, IBM, and Dupont) are in the race to get a piece of this pie, there are four big publishing houses that have virtually monopolized the industry: Harcourt Brace, Houghton Mifflin, Pearson, and McGraw-Hill. As Stephan Metcalf (2002) notes, "the so-called Big Three—McGraw-Hill, Houghton-Mifflin and Harcourt—[were] all identified as 'Bush stocks' by Wall Street analysts in the wake of the 2000 election."

It is interesting to note that three of the four giants are internationals, and given the conservative pitch on global competitiveness and national security, other than for financial gain, it's not clear why such companies would have any vested interest in improving the education of students in the United States.

Harcourt Brace is owned by London-based publisher Reed Elsevier. With contracts in eighteen states, the company is known in the testing industry for the SAT-9, STAR, and Wechsler Intelligence Scale tests, and as mentioned earlier, it played a significant role in drafting Texas's

TAAS exam. Willing to go to almost any length to maximize its profits in Texas, Harcourt pitched its course materials as being published by "the same company that helps to write the TAAS tests" (Bacon, 2000). What is unethical and self-serving about this practice is that any use of prep materials that developmentally coincide with the content of the exam compromises the validity of the scores (Gluckman, 2002). In other words, prepping of this kind is a subtle form of cheating; but for Harcourt, student success on the exam means guaranteed contracts.

By the end of the millennium, when the testing movement was really heating up, Harcourt's educational division was pleased to inform its shareholders of an almost 30 percent increase in sales. The international conglomerate, which also owns Holt, Rinehart and Winston, boasts annual revenues of over five billion dollars.

The Houghton Mifflin Company, formerly owned by the French corporation Vivendi Universal, is currently the property of the international conglomerate HM Publishing Corp. With contracts in thirteen states it is primarily in the business of selling textbooks and instructional technology. Its testing division, Riverside, is known for the Iowa Test of Basic Skills (ITBS), the Metropolitan Achievement Test, and two assessments that comply with the "Reading First" mandates of NCLB—the Gates MacGinitie Reading Test and the Basic Early Assessment of Reading (BEAR).

Houghton Mifflin also gained from the early race to nationalize standards as its testing division's profits grew by almost 18 percent in 1999. With a keen understanding of how information processing is a key component of the standards market, in 2003 Houghton Mifflin purchased Edusoft, a profitable company that specializes in data storage and online tests. The conglomerate now boasts more than one billion dollars in annual sales.

Pearson Education is under the flagship of London-based Pearson Publishing, which also owns Penguin Books, *The Economist*, Simon & Schuster, and *The Financial Times*. For years it has maintained a working relationship with NAEP, scoring exams, and it is currently the leading scorer of standardized tests in the country. To assist and expand these efforts, in 2000, Pearson bought National Computer Systems at a cost of $2.5 billion.

When Congress allocated three billion dollars per year for teacher training in its reauthorization of the Elementary and Secondary Education Act in 2001, Pearson decided to take a big bite out of that market share by acquiring National Evaluation Systems, Inc. (NES), which produces assessments for teacher certification. A key characteristic of the new "highly qualified teacher," according to NCLB, is his or her ability to pass a subject matter test administered by the state. Reducing teacher expertise to a fixed body of content knowledge, teachers are expected to meet an extremely narrow range of skill requirements under the new policy. Any concern with pedagogy—not what we learn, but how we learn it—has virtually disappeared.

Teacher burnout, a serious side effect of this era of standards and accountability, is actually a virtue in a profit-driven industry as new teachers require certification. As Pearson notes in its 2006 press release:

> There are approximately three million public school teachers in the U.S. Approximately 2.5 million new teachers will need to be hired in the current decade, as 700,000 current teachers retire and 1.8 million are expected to leave the profession prior to retirement. On average, nearly 6% of the teacher workforce does not return for each new academic year and half of all new teachers leave the profession within five years.

Pearson currently has long-term contracts with more than twenty states and its 2005 profits were up 29 percent (Forbes.com, 2006). Since the implementation of NCLB, their sales on assessments alone are up more than 20 percent. In its 2005 "Performance Report" under the subtitle "Continued Investment for Future Growth," Pearson reassures its shareholders how it will greatly increase its profits from U.S. schools by a steady investment in school publishing, basal curriculum programs for reading, science, and social studies, and school testing (where it already maintains contracts in Texas, Virginia, Michigan, and Minnesota that have a lifetime value of seven hundred million dollars). It also mentions the creation of Pearson Achievement Solutions that targets the growing market for teacher professional development and integrated school solutions. It's interesting to note that there isn't a single word in the report about the academic achievement of students.

With 280 offices in 40 countries, McGraw-Hill is a major player in the publishing world and among its treasured assets are Standard & Poor's, *Business Week*, four television stations, and the online data analysis company Turn Leaf Solutions. A simple look at this New York-based company's website and one can see how profits have consistently soared since the advent of the standardization craze (McGraw-Hill, 2005, 2006).

With contracts in twenty-three states, McGraw-Hill's Terra Nova, CTBS, and California Achievement tests are the most lucrative of its assessment instruments. Trying to expand its $1.4 billion textbook sales, "McGraw-Hill lobbyists used the statewide results on their own California Achievement Tests to convince the state legislature that California schools needed the McGraw-Hill Open Court and Reading Mastery program to improve students reading performance" (Clark, 2004). Expected to teach at least part of the day from the McGraw-Hill Open Court materials,

> according to Ben Visnick, [president of a local teachers union in California], "School district employees and instructional facilitators—we call them Open Court police—inspect the classrooms to verify that the right posters are on the walls and they want everyone in the district on the same page every day." (As cited in Clark, 2004)

While the California Department of Education guidelines prohibit the use of test-prep materials written for a specific test, the practice is common nonetheless (Gluckman, 2002).

In the world of the Fortune 500 (2006), McGraw-Hill comes in at 359 with over six billion dollars in annual revenues. But perhaps what's most interesting about this corporation is its deep connection to the Bush dynasty. In his article in *The Nation*, "Reading between the Lines," Stephen Metcalf (2002) lays out the depth of this nepotism, describing how the two families have been chummy since the 1930s when they vacationed together in an exclusive area in Florida. He is worth quoting at length here:

> Harold McGraw Jr. sits on the national grant advisory and founding board of the Barbara Bush Foundation for Family Literacy. McGraw in turn received the highest literacy award from President Bush in the early 1990s...The McGraw Foundation awarded current Bush Education Secretary Rod Paige its highest educator's award while Paige was Houston's school chief; Paige, in turn, was the keynote speaker at McGraw-Hill's "government initiatives" conference last spring. Harold McGraw III was selected as a member of President George W. Bush's transition advisory team...An ex-chief of staff for Barbara Bush is returning to work for Laura Bush in the White House—after a stint with McGraw-Hill as a media relations executive. John Negroponte left his position as McGraw-Hill's executive vice president for global markets to become Bush's ambassador to the United Nations.

And of course, under Bush Jr., Negroponte would go on to be U.S. ambassador to Iraq (2004–05), and is now the director of national intelligence. The word "intelligence" here has two frightening implications: a scary thought in terms of national security given the well-documented horrors that Negroponte was involved in while trying to subvert the growth of democracy in Latin America when he was U.S. ambassador of Honduras (1981–85) under Reagan; and a scary thought in terms of what our children learn in schools that are under the influence of a standards regime that works diligently to engineer history as it sees fit; much in the way that Negroponte himself worked to keep his actions in Latin America from becoming public knowledge.

McGraw III, on the other hand, as part of a group of "education leaders," was invited to speak at the White House by George W. Bush on his first day in office.

Bush Jr. and McGraw-Hill have been partners in crime before. As governor, Bush joined forces with the publishing giant in order to pitch their proposed phonics-based literacy program to the Texas Education Agency, which was trying to establish a statewide reading curriculum. As is to be expected, these experts tooted their own horns in front of TEA, calling for a reading program that was right in tune with a slew of new textbooks and materials from McGraw-Hill—a market that the company easily cornered with the support of the governor (Clark, 2004).

Conservatives insist, ad nauseam, that "scientifically-based research" inform and sustain the nation's educational practices, policies, and goals. However, the empirical studies that are used to buttress the Republican agenda, under close scrutiny, are easily stripped of any legitimacy (Allington, 2002; Coles, 2000). The corporate bodies and the well-funded private think tanks that produce much of the research and literature to support these causes have an obvious ideologically specific take on these issues, one that is widely supported by mainstream media whose ownership have similar interests.

A recent and striking example of how data can be manipulated, packaged, and presented as "scientific research" is the official report signed and circulated by the congressionally appointed National Reading Panel (NRP). The report, which informed Bush's multibillion dollar Reading First literacy program campaign, is replete with inconsistencies, methodological flaws, and blatant biases (Allington, 2002; Coles, 2000). For starters, Bush's educational advisor when he was the governor of Texas, G. Reid Lyon, headed the NRP. A staunch phonics advocate, Lyon hand-selected the panel and made certain that virtually all of the participants shared his views. Curiously, there was only one reading teacher on the NRP. However, by the end of the group's investigation into effective literacy practices, she refused to sign the panel's final report, maintaining that it was a manipulation of data, and that the cohort failed to examine important research that did not corroborate its desired findings. As Metcalf (2002) reveals:

> Widmeyer [the public relations firm hired by the government to promote the panel's work] had represented McGraw-Hill's flagship literacy product Open Court during the Texas literacy drive, and now it counts McGraw-Hill and the Business Roundtable among its most prominent clients. "They wrote the introduction to the final report," says NRP member Joanne Yatvin. "And they wrote the summary, and prepared the video, and did the press releases."

McGraw-Hill has been laughing all the way to the bank ever since, tapping into the six billion dollars that the president has set aside to fund his "literacy" campaign. And guess who's been recruited to hold the federal purse strings: Christopher Doherty, the guy who spearheaded the move to bring McGraw-Hill's DISTAR (Direct Instruction System for Teaching Arithmetic and Reading) to public schools in Baltimore (Metcalf, 2002).

In the world of crony capitalism, these kinds of deals are made all the time. Look at how Bill Bennett, the former secretary of education under Reagan (1985–88), and drug czar under Bush Sr. (1989–90), has been cashing in lately. Bennett's online home/school company K-12 Inc. recently received four million dollars in grant money from the U.S. Department of Education. The funds are intended for an online charter school in Arkansas—Arkansas Virtual Academy (The Homeschool Place.com, 2006).

Not only is it morally questionable how Bennett's for-profit business came about getting the grant, especially since other programs that had been turned down had better independent reviews (Ohanian, 2006), but it is also disconcerting that federal funds are being diverted away from public schools in order to subsidize education for homeschool students. Sure, NCLB has set aside money for its Voluntary Public School Choice Program with the expressed purpose of giving students a chance at a better education; however, only 25 percent of the students who have participated in K-12 Inc.'s program are from public schools (Ohanian, 2006).

Meanwhile, Bennett has been working his inside contacts to cut deals in other states around the country. Though he extols universal morality in his *Book of Virtues* (1993), it is important to remember that this is a guy who has a multimillion dollar gambling addiction, and who on his syndicated radio talk show expressed to a caller: "If you wanted to reduce crime, you could—if that were your sole purpose—you could abort every black baby in this country and your crime rate would go down" (CNN.com, 2005). This is a guy who doesn't even support public education and yet uses taxpayers' money to finance his own business ventures. As Intel director and former FCC chair Reed Hundt reveals about the former secretary of education:

> I asked Bill Bennett to visit my office so that I could ask him for help in seeking legislation that would pay for internet access in all classrooms and libraries in the country . . . He told me he would not help, because he did not want public schools to obtain new funding, new capability, new tools for success. He wanted them, he said, to fail so that they could be replaced with vouchers, charter schools, religious schools, and other forms of private education. Well, I thought, at least he's candid about his true views. (As cited in Hoffman, 2005)

It's also important to remember that Bennett used start-up money from Knowledge Universe to get K-12 Inc. up and running. Knowledge Universe is owned by Michael Millikin, the "Junk Bond King" who ripped off investors for billions of dollars in the 1980s and consequently spent a couple of years behind bars—actually, there were no bars where Millikin did time. And when he was done with his sentence, a period during which he learned how to prepare traditional French dishes from the in-house chef, he was allowed to keep over two billion dollars that he had accumulated from his criminal escapades. It certainly doesn't seem virtuous for Bennett to use blood money for any purpose, let alone to profit from the very taxpaying public from which it was ultimately stolen.

There is ample room for nepotism in state and federal politics, especially when it comes to family connections. As revealed in the *USA Today* article "Congress Full of Fortunate Sons—and Other Relatives" (Lawrence, 2006), there are more than fifty U.S. senators and representatives that are closely related to governors and other members of Congress. But what better position is there to be in when it comes to old-boy networks than to be the grandson of a former senator, the son of a former CIA director and president of the United States, the brother of the incumbent governor of Florida, and the brother of the former governor of Texas—now president of the country.

Neil Bush, the president's youngest brother, in the spirit of NCLB, is pushing to sell to states around the country online, multimedia educational products and test-prep software produced by his Austin-based company Ignite Inc. Founded in 1999, Ignite set its sights on Florida public schools where the company has been pitching its products as helping students prepare for the Florida Comprehensive

Assessment Test (FCAT). While the Florida Education Association has expressed some concern about a potential conflict of interest in developing a business relationship with Neil Bush, given that his brother is the governor, the youngest sibling adamantly denies any discussion about his business affairs with either of his two brothers.

Ignite Inc. certainly has an advisory board fit for the job of soliciting state and federal monies:

> Ignite has loaded its advisory committees with Bush loyalists, assuring the company a sympathetic ear in Washington. According to the company, its big-name consultants include Bill Brock, a former senator from Tennessee who chaired the Republican National Committee; Bob Stearns, a Houston investor appointed to a Texas technology board by George W.; Peter Su, a former campaign adviser of the president, and two executives from Bessemer Trust, an exclusive investment firm that manages a portfolio for Neil's dad. (Scherer, 2001)

The reason that most people don't know much about the youngest Bush heir is largely because his political career died in the late 1980s—during his father's tour as president—when he was the acting director of Silverado Savings and Loan in Colorado, a company that went down with the S&L ship. Neil Bush's scandalous behavior cost taxpayers over one billion dollars.

What Can Educators Do?

Progressive educators and community activists desperately need to continue to do the important work of informing the public about the privatization of their schools. Ultimately, the long-term goal is to have a critically informed public vote out of office representatives that are sacrificing children to the corporate bottom line with prepackaged teacher-proof curricula, standardized tests, and accountability schemes. But in the meantime, there are things that teacher education programs and practitioners can be doing to work toward redemocratizing public schools and creating civic-minded students and a vibrant public sphere.

First and foremost, teacher education programs need to be structured in a way that helps apprentice students into conducting critical inquiry. Not to be confused with what's traditionally thought of as higher order thinking skills, *critical* in this sense implies being able to understand, analyze, pose questions, and affect the sociopolitical and economic realities that shape our lives (Freire, 1970). It is important to emphasize that developing critical consciousness—or what I often refer to as presence of mind—isn't an exercise to get people to think in a certain way ideologically, rather it is intended to get us to explore in greater depth the issues and relations of power that affect the world around us (Leistyna, 1999, 2002). Unlike the president's agenda that deskills and disempowers educators, as revealed in the following comment: "*When I picked the Secretary of Education, I wasn't interested in a theorists...*" (White House ceremony honoring the 2002 National Teacher of the Year), civic-minded teacher education programs need to create the conditions within which people can think for themselves. Teachers shouldn't be mere practitioners who are trained to jump when they are told to jump or technicians whose only purpose is to read materials to the class and handout and collect high-stakes tests. They need to be intellectuals and professionals that can make sense of—that is, theorize—the world around them and make informed, critical, and ethical decisions. They need to engage in praxis—the ongoing relationship between theory and practice/reflection and action, and teacher education programs can play a significant role in this development.

In the spirit of theory and theorizing, future educators should study in detail the standards and assessment movement during their graduate work. They need to learn about the historical developments therein and the policies that inform such practices, and the people behind them. They should be examining the content formulation and evaluation procedures of high-stakes tests and the empirical research that explores this terrain. Teacher education programs can arm students with the skills necessary to look into the political economy of this industry and the synergy that exists within among the private and public sector. They can effectively apprentice students into deconstructing the theoretical frameworks that undergird the practice itself. Educators need to ask themselves on a regular basis: Why exactly are we doing this? This is surely better than what the president does in his daily duties—as he revealed onboard Air Force One in 2003: "I'm also not very analytical. You know, I don't spend a lot of time thinking about myself, about why I do things."

Teacher education programs also need to continue to do extensive research and help put into practice more complex and nuanced forms of evaluation such as portfolio assessment. Private schools—where the elite, ironically including the most vociferous proponents of the testing industry, send their own children—don't have high-stakes tests, don't have to abide by the rules and regulations of NCLB, and have far more in-depth evaluative processes.[4]

As teachers are activists, whether or not they want to be—given that they have an effect on people everyday—part of their graduate experience should help them become more effective agents of change. As part of this process, teacher education programs should include case-study analysis of movements that both support and reject high-stakes testing. There are a plethora of case studies that could be used for this purpose. In support of testing, they could analyze the work of the Business Rountable. On the flip side, there is much to learn from the actions of the Coalition for Authentic Reform in Education (CARE), Californians for Justice (CFJ), The Mexican American Legal Defense and Education Fund, and other such organizations and activists groups around the country that have been effectively working to mobilize the community against the kinds of social injustice and profiteering that the privatized high-stakes movement engenders. There is also a plethora of teacher and student movements that are worth studying, such as the Student Coalition for Alternatives to the MCAS (SCAM).

The intent of exploring this kind of activism is not to offer up a recipe book to be followed to the last grain of salt; rather, it is a way to inspire critical appropriation and the reinvention of social change as these struggles offer a theoretical, empirical, and practical springboard for future efforts.

While the title No Child Left Behind connotes fairness, compassion, and equity, and the instigators of testing mania promise academic and professional success for our children, these political campaigns virtually disregard why inequities exist in the first place. As advocates of the corporate model of schooling hide behind notions of science, objectivity, and universal knowledge, what is largely missing from national debates and federal and state policies, and what should be central to any good teacher education program, is a recognition and analysis of how racism, the structures of social class, and other oppressive and malignant ideologies inform actual educational practices and institutional conditions. These factors play a much more significant role in students' academic achievement than whether or not they have access to abstract content, a monolingual setting, and constant evaluation. Instead of honestly confronting these issues, conservatives readily blame progressive educational programs, democratic social policies, and organized labor for the country's problems.

While standards are a way of limiting the power of labor unions, teacher education programs can be designed to expose students to the extensive history of organized labor's struggles to improve education both domestically and internationally. It is important to note that with improvements in technology and computers, standardized testing has become a global industry and thus a global problem (Rees, 2000). Future educators need to learn how to work with teacher unions to collectively organize, and they need to be able to effectively mobilize these organizations to act on behalf of the community. For example, how does one go about pressuring the federal government to fund independent watchdogs to help keep an eye on assessment practices in the United States?

But K-12 students also need to learn how to be more effective agents of change and need to be encouraged to engage in praxis themselves. If testing is such an important part of society as advocates of this industry claim, then students should also know more about the practice—not just what's on the test and how to take it, but how the tests are generated and by whom. This opens the possibility of an interdisciplinary approach to learning where subjects such as history, math, science, and social studies are part of analyzing the industry and the tests themselves. In the spirit of critical pedagogy, teachers should also teach to the test. But by this I mean that they should engage the students in ideological analysis of the knowledge that they are being exposed to. It is important for young people to take a critical look at, for example, the history lessons that are taught in schools whose curricula and textbooks are generated by the current standards regime. Whose stories are told, how are they told, what is being excluded in such representations, what purpose and whose interests do such stories serve? Again, teacher education programs could play an important role in nurturing this kind of pedagogical practice.

In this era of massive corporate corruption of the likes of Enron, World Com, Adelphia, Wal-Mart, and Tyco, where unscrupulous political figures such as Tom Delay, Ralph Reed, Jack Abramoff, and David Safavian are readily caught engaging in criminal activity with the private sector, it simply makes no sense to put our nation's youth in the "trusting hands" of corporations. The results that matter for private industry are financial gains achieved by selling materials on a grand scale, and by guaranteeing that schools produce an uncritical mass of a low and semi-skilled labor that is in high demand in our now postindustrial service-oriented economy. Why else would they support policies that offer a mechanical and mediocre curriculum, and that force kids out of school prematurely, guaranteeing that they don't get a high school diploma and thus access to college?

Leaving public education in the hands of for-profit corporations would be like letting HMOs and pharmaceutical and insurance companies tell doctors what to do; it would be like letting Exxon/Mobil and construction companies such as Halliburton make decisions about whether or not this nation goes to war. Unfortunately, this is precisely what the neoliberal approach to governing and decision-making has done. It's about time we redemocractize this society and if public schools are to play a key role in this change, we need to be honest about what these institutions are designed to do. My favorite Bush misspeak is:

> The public education system in America is one of the most important foundations of our democracy. After all, it is where children from all over America learn to be responsible citizens, and learn to have the skills necessary to take advantage of our fantastic opportunistic society. (May 1, 2002)

While this statement is an obvious slip of the tongue, and, sadly, a sign of a semi-literate man, in reality what it says isn't too far from the truth. Part of being a democracy means having public institutions that are designed to create a vibrant and critical citizenry. However, within the grips of unbridled capitalism, private interests have worked to create a get-what-you-can-for-yourself mentality. To really create responsible citizens it means that a society has to collectively participate in maintaining healthy public institutions and restraining those opportunistic forces that have an undying love for profit, but such contempt for public life.

Notes

1. For a much more detailed version of the early days of this corporate story, see Suchak (2006). This research informs much of this section of my chapter and should be read in its entirety.
2. Like the "Vallas Miracle" in Chicago and the "New York City Miracle."
3. This reference is taken from the White House webpage at http://www.whitehouse.gov/government/paige-bio.html.
4. Pepi Leistyna. (2007). Corporate testing: standards, profits, and the demise of the public sphere. *Teacher Education Quarterly*, 34(2), 59.

References

Allington, R. (2002). *Big Brother and the National Reading Curriculum: How Ideology Trumped Evidence*. Portsmouth, NH: Heinemann.

Bacon, D. (2000). School testing: Good for textbook publishers, bad for students. *Pacific News Service*. Available at http://alternet.org/story/39/.

Bacon, D. (2001). The trouble with the test. *Students*. Available at http://dbacon.igc.org/Students/04TheTest.htm.

Bennett, W. (1993). *Book of Virtues: A Treasury of Great Moral Stories*. New York: Simon & Schuster.

Clark, B. (2004). Leaving children behind: Exam privatization threatens public schools. CorpWatch. Available at http://www.corpwatch.org/article.php?id=11543.

CNN Presents. (2005). High stakes. Cable News Network Inc. Lanham, MD: CNN.

CNN.com (2005). Bennett under fire for comments on blacks, Crime. Availableathttp://64.233.161.104/search?q=cache:MEhjtC3_OI4J:www.cnn.com/2005/POLITICS/09/30/bennett.comments/+Bill+Bennett,+the+former+Secretary+of+Education&hl=en&gl=us&ct=clnk&cd=1.

Coles, G. (2000). *Misreading Reading: The Bad Science that Hurts Children*. Portsmouth, NH: Heinemann.

Committee on Education and the Workforce: U.S. House of Representatives (1999). Statement of Edward B. Rust, Jr., Chairman and CEO of State Farm Insurance Companies; Committee on Education and the Workforce: U.S. House of Representatives on "Business Issues in elementary and Secondary Education." Available at http://www.house.gov/ed_workforce/hearings/106th/fc/esea7199/rust.htm.

FairTest: The National Center for Fair and Open Testing (2006). The case against high-stakes testing. Available at http://www.fairtest.org/arn/caseagainst.html.

Forbes.com (2006). Round-up Pearson FY profit climbs. Available at http://www.forbes.com/markets/feeds/afx/2006/02/27/afx2555308.html.

Fortune. (2006). Fortune 500. Available at http://money.cnn.com/magazines/fortune/fortune500/.

France, M., and Osterland, A. (1999). State Farm: What's happening to the good neighbor? Judges and juries have found the insurer guilty of serious misconduct. *Business Week Online*. Available at http://www.businessweek.com/1999/99_45/b3654189.htm.

Freire, P. (1970). *Pedagogy of the Oppressed*. London: Penguin.

Frontline. (2006). The testing industry's big four. Available at http://www.pbs.org/wgbh/pages/frontline/shows/schools/testing/companies.html.

Gluckman, A. (2002). Testing...testing...one, two three: The commercial side of the standardized-testing boom. *Dollars & Sense: The Magazine of Economic Justice*. Available at http://www.dollarsandsense.org/archives/2002/0102gluckman.html.

Haney, W. (2000). The Texas miracle in education: Missing students and other mirages. *Education Policy Analysis Archives*. Available at http://epaa.asu.edu/epaa/v8n41/ and http://epaa.asu.edu/epaa/v8n41/part5.htm.

Hoffman, T. (2005). The Candid Bennett. *eSchool News: Ed-Tech Insider*. Available at http://www.eschoolnews.com/eti/2005/10/001190.php.

Horn, J. (2005). Hauling away the Federal Treasury. Available at http://schoolsmatter.blogspot.com/2005/09/hauling-away-federal-treasury.html.

Humes, E. (2003). Leave no test behind. Available at http://www.edwardhumes.com/archives/00000021.shtml.

Kohn, A. (2000). *The Case against Standardized Testing: Raising the Scores, Ruining the Schools.* Portsmouth, NH: Heinemann.

Kohn, A., and Shannon, P. (2002). *Education, Inc.* Portsmouth, NH: Heinemann.

Lawrence, J. (2006). Congress full of fortunate sons—and other relatives. *USA Today.* Available at http://www.usatoday.com/news/washington/2006-08-07-relatives-cover_x.htm.

Leistyna, P. (1999). *Presence of Mind: Education and the Politics of Deception.* Boulder, CO: Westview.

———. (2002). *Defining and Designing Multiculturalism.* Albany, NY: SUNY.

McGraw-Hill. (2005). Report. Available at http://investor.mcgraw-hill.com/phoenix.zhtml?c=96562&p=irol-newsArticle&ID=700914&highlight=.

———. (2006). Report. Available at http://investor.mcgraw-hill.com/phoenix.zhtml?c=96562&p=irol-newsArticle&ID=885916&highlight=.

McNeil, L. (2000). *Contradictions of School Reform: Educational Costs of Standardized Testing.* New York: RoutledgeFalmer.

Metcalf, S. (2002). Reading between the Lines. *The Nation.* Available at http://www.fairtest.org/nattest/Nation%20piece%20bush%20links.html.

Mickelson, R.A. (2006). International Business Machinations: A case study of corporate involvement in local educational reform. *Teachers College Record.* Available at http://72.14.209.104/search?q=cache:zmhIeO-xRfUJ:www.uncc.edu/rmicklsn/images/corporate.pdf+New+American+Schools+Development+Corporation,+initially+funded+by+IBM+CEO+Louis+Gerstner&hl=en&gl=us&ct=clnk&cd=1.

Miner, B. (2005). Keeping public schools public: Testing companies mine for gold. *Rethinking Schools.* Available at http://www.rethinkingschools.org/archive/19_02/test192.shtml.

Molnar, A. (2005). *School Commercialism.* New York: Routledge.

Molnar, A., Garcia, D., Bartlett, M., and O'Neill, A. (2006). Profiles of for-profit education management organizations 2005–2006. *Education Policy Research Unit.* Available at http://www.asu.edu/educ/epsl/CERU/Documents/EPSL-0605–104-CERU.pdf.

Ohanian, S. (2006). Grant to Bennett's K12 Inc. Challenged. Available at http://www.susanohanian.org/atrocity_fetch.php?id=2910.

Olson, D. (2005). Testing companies score big profits. Minnesota Public Radio. Online at http://news.minnesota.publicradio.org/features/2005/04/18_olsond_testingagain/. Accessed November 5, 2007.

Pearson. (2005). Report. Available at www.pearson.com/index.cfm?pageid=151.

———. (2006). Press Release: Pearson Enters Certification Market, Acquires National Evaluation Systems, America's Leading Teacher Certification

Testing Company. Available at http://www.pearsoned.com/pr_2006/042506.htm.

Rees, M. (2000). Educational testing as a global industry. World Socialist Website. Available at http://www.wsws.org/articles/2000/aug2000/test-a22.shtml.

Sacks, P. (2001). *Standardized Minds: The High Price of America's Testing Culture and What We Can Do to Change It*. Boulder, CO: Perseus.

Saltman, K. (2000). *Collateral Damage*. Boulder, CO: Rowman & Littlefield.

———. (2005). *The Edison Schools: Corporate Schooling and the Assault on Public Education*. New York: Routledge.

Scherer, M. (2001). That other Bush boy. *Mother Jones*. Available at http://www.motherjones.com/news/outfront/2001/05/neilbush.html.

Sixty Minutes II. (2004). The "Texas Miracle." CBS News. Available at www.cbsnews.com/stories/2004/01/06/60II/main591676.shtml.

Sturrock, C. (2006). States Distort School Test Scores, Researchers Say. Critics Say California among Those that Lower Standards for No Child Left Behind. *Chronicle*, June 30. Available at http://www.sfgate.com/cgi-bin/article.cgi?file=/c/a/2006/06/30/MNG28JN9RC1.DTL.

Suchak, B. (2006). Standardized testing: High-Stakes for students and for corporate bottom lines. Available at http://www.nomoretests.com/insider.htm.

TC Reports. (2000). High-stakes testing and its effects on education. Teachers College. Available at http://www.tc.columbia.edu/news/article.htm?id=3811.

Texas Education Agency. (2000). Press Release. Available at http://72.14.207.104/search?q=cache:x6NRKCIx-BkJ:www.tea.state.tx.us/press/pr000407.html+students+in+texas+take+TAAS&hl=en&gl=us&ct=clnk&cd=1.

Texas NAACP. (1995). TAAS Resolution Commitment. Available at http://72.14.207.104/search?q=cache:zzHwpR6B1bIJ:www.texasnaacp.org/taasrc.htm+TAAS+exam,+english+and+spanish&hl=en&gl=us&ct=clnk&cd=10.

The Homeschool Place.com (2006). X U.S. Secretary of Education Bill Bennett and President Bush's Brother Neil Bush. Available at http://www.edu-cyberpg.com/Culdesac/billbennett.html.

Chapter 4

Revolutionary Critical Pedagogy: The Struggle against the Oppression of Neoliberalism—A Conversation with Peter McLaren

Sebastjan Leban and Peter McLaren

Sebastjan Leban: On your webpage you state that critical pedagogy, which you support and practice, advocates nonviolent dissent, the development of a philosophy of praxis guided by a Marxist humanism, the study of revolutionary social movements and thought, and the struggle for socialist democracy, which is diametrically opposite to the current neoliberal democracy. Can we say that you as a critical educator are basically leading a fight against neoliberal global capitalism valorization of education?

Peter McLaren: Yes, that would be a very fair description. I am not alone, clearly, in this struggle. There are others within universities who work internationally toward the same goal. But it is also fair to say that in the United States there are very few of us in the field of education and the future in this regard looks irrepressibly bleak. Well, let me put it this way. To their ongoing credit, there are those who are quite capable of engaging in a rigorous analysis of mounting sophisticated and nuanced attacks on the scoundrels and hypocrites of the Bush administration, the gangster capitalists and political opportunists, the feckless cabal of Christian-right "profamily activists" who exercise their racism by warning about the coming "demographic winter" facing the United States unless the white population produce enough babies to achieve "replacement-level fertility," and on the evangelical economic zealots who call out for a renewed assault on the poor through neoliberal economic, political, and social directives and principles. But these critics of the wretched havoc wrought by neoliberalism do not at the same time identify an alternative—at least one that is couched beyond very safe and I would say largely empty liberal pluralist principles. Until about the mid-1990s, I found myself in the same dilemma. For me, the struggle was focused on democratizing the public sphere. But since that time I have been a staunch advocate of education as a means to further socialism, that is, to bring about a world outside capital's valorization process or, put another way, outside labor's value form. I have described capitalism and democracy as two thieves planning a joint robbery and simultaneously attempting to steal the spoils from each other.

I have been part of a movement to build a radical humanistic socialism—in part by de-writing socialism as a thing of the past and rewriting critical pedagogy as a struggle for a postcapitalist alternative—and in doing so I have taken the position that socialism and socialist principles are not dead letters, but open pages in the book of social and economic justice yet to be written or rewritten by people struggling to transform our capitalist prehistory, and to build a truly egalitarian social order where, as Marx put it, the real history of humanity can begin. We can do this in a number of ways but I have been concentrating mainly but not exclusively on ideology critique, de-naturalizing what is assumed to be unchangeable, de-reifying human agency, and de-objectifying the commodity culture of contemporary capitalism. I have been trying to discourage progressive educators from a sole reliance on a politics of human rights antiseptically cleaved from the issue of economic rights and have been trying to unburden cultural studies of its textuality of the negative, what Marxist professor Teresa Ebert calls a "site of meaningfulness without meaning and thoughtful unthoughtfulness" that presumably arrived on the wings of the Angel of History to save us from the old bearded devil: Karl Marx. With the advent of the linguistic turn in the arts and social sciences—a time regrettably, where class struggle was rewritten in the aerosol terminology of the politics of difference, and difference were treated as difference within itself (how difference is different from itself)—Marxism was a popular target among progressive academics. But replacing class struggle with the politics of "difference" and "diversity" flattens out and empties the whole structure of antagonism or ensemble of relations of opposition within the structured hierarchy of capitalist social relations. Social relations of oppression are, in this case, dissolved into difference within or between two differences—into relations of supplementarity—rather than highlighting labor relations or struggles between workers and the capitalist class. Ebert has written in great detail on this in a white heat but also with an attempt at clarity.

And yes, of course, I have at the same time been challenging what Quijano calls "the coloniality of power" (I very much admire the work of Quijano, Mignolo, and Grosfoguel although I have some difficulty with some aspects of their critique of Marxism). Educators, especially, need to get beyond the manufactured fear and the hysterical rhetoric, peddled by what we call the corporate-state-military-media complex (or simply, the "power complex"), and instead seek a deeper means of challenging repressive and violent social structures. In some instances, we might slow down and reverse the current trend among legislative and policy-making bodies and political leaders who contribute mightily to the dreck and the moral refuse that has come to define the current war against the poor within the United States and the struggle against the working class by the transnationalist capitalist class. Since 1987, I have been visiting radical educators, student groups, workers, philosophers, counterculturalists, contrarians, culture brokers, and pedagogical tastemakers internationally (most recently in Finland, Portugal, Greece, Venezuela, Brazil, and Cuba), attempting to conscript their messages into a larger, transnational drumbeat that will help to entrain an activist movement toward a postcapitalist alternative.

What has been different about my work over the last decade is that it delves deeper into the terrain of Marxist theory and with more exigency and urgency, in an attempt to create spaces/places in different scales and registers where students can apprise themselves of the opportunity to resist more fully the geopolitics of imperialism and comprehend how new social relationships can be wrought that can supercede those given birth by the United States' underbelly of violence—a poisonous underbelly festering inside a hypocritical miasma of couth that floats everywhere and penetrates the very structure of our consciousness through electronic orifices that make up the neoliberal sensoria of propaganda—propaganda that is imbibed at last partially by a mystified duped citizenry under the aroma of "democracy." All of this is part and parcel of the geopolitics of imperialism that largely defines U.S. foreign and domestic policy—all of which, of course, impacts how we both view and develop our role as educated citizens (in my case, a citizen of the world as I am against most forms of nationalism) and critical cosmopolites. The gangrene-ridden wound in the soul of the country won't be healed by Obama or McCain (certainly not McCain who is a total nutcase). The issue goes beyond the United States itself. It has to do with the transnational capitalist class.

But the United States certainly plays a major role. In light of the Bush administration's "humanitarian" invasion of Iraq, and other U.S. war crimes too numerous to mention, its current war on the poor, its savage repression of twelve million immigrant workers, and its involvement in overthrowing democratically elected regimes worldwide, we must detach from the term democracy the connotations of equality before the law, free speech, right of association, universal suffrage, and self-rule with which it has been saddled over the decades and recognize it as a vile condition that ensures the involuntary servitude of wage labor, the racial and gendered division of labor, and the plundering of natural resources by the imperial powers.

The once grand refusal of critical pedagogy to reproduce dominant ideologies and practices inherent in capitalist schooling and the wider context of globalized capitalism and instead to embrace the possibility of decolonizing the conceptual, philosophical, epistemological, and cultural dimensions of learning has been expurgated by the flat-lined antipolitics of postmodernism. My work has set itself up in opposition to this fashionable apostasy undertaken by what I once termed the avant-garde "hellions of the seminar room."

Gore Vidal once presciently noted that the U.S. government prefers that "public money go not to the people but to big business. The result is a unique society in which we have free enterprise for the poor and socialism for the rich" and the truth of this statement is no more evident than in the recent nationalization of Fannie and Freddie where you can see clearly that the United States is a country where there exists socialism for the rich and privatization for the poor, all basking in what Nouriel Roubini calls "the glory of unfettered Wild West laissez-faire jungle capitalism" that has allowed the biggest debt bubble in history to fester without any control, causing the biggest financial crisis since the Great Depression. Indeed, socialism is only condemned when it profits the poor and the powerless and threatens the rich. But capitalists are quick to embrace a socialism for the rich—which really is what neoliberal capitalism is all about. But of course, it's not real socialism but a form of state capitalism. Which is why today we have democracy for the rich while the poor are cast into quasi-feudal steampunk landscapes of dog-eat-dog despair. Those whose labor is exploited in the production of social wealth—that is, the wage and salaried class—are now bearing most of the burden of the current economic crisis in the United States.

S.L.: In the interview "Pedagogy and Praxis in the Age of Empire" (also the title of one of your books) published in the fall of 2007 you argue that revolutionary critical pedagogy operates from an understanding that the basis of education is political and that we have to create a space where students can be given resources to imagine a different world outside the capitalism's law of value. Could you describe what space in particular you have in mind? Can you define the moment of the revolutionary in critical pedagogy?

P.M.: The moment of the revolutionary. I like that term. I suppose that there are as many revolutionary moments as there are critical educators. Let me wind up to your answer by providing some theoretical context. As I expressed this dilemma and challenge in an article recently, while it is certainly true, as many post-structuralists unguardedly claim, that we are semiotically situated in hermeneutic horizons, in gendered and racialized positionalities riven by power-sensitive and power-expansive relations of symmetrical privilege, and in social space aligned and vectored geopolitically and cross-hatched socioculturally, it is also true that the totalizing power of capital has created an overarching matrix of exploitation in which all of these antagonisms have been accorded value in relation to the sale of human labor power in the global marketplace where, like force-fed swine who are made blind and crippled in preparation for mass consumption, men and women are led to the slaughterhouse of capital hoisted on hooks of poverty and debt. By this I meant that we certainly should not refrain from exploring and celebrating our ethnic heterogeneity and heterodox temporalities that power our subjectivity. I am not against this, or related issues such as building border identities that escape the lineaments of Eurocentric epistemes. This is all fine and good. But let's not forget that the totalizing power of capital creates constitutive limitations in which subjectivities are formed. This, I have argued, can be seen as a form of controlled consent made possible by the production of social amnesia both produced and enforced by the corporate media, and the deep psychology that turns the engines of mass propaganda disguised as a free marketplace of ideas (where the only free cheese available is in the mousetrap). Democracy has become synonymous with profit-making, requiring a rollback of trade union power and a generalized hollowing out of social democracy, not by military dictatorship but by an endless stream of maledictions and execrations against leftist movements and Marxist analyses that deal with the totality of capitalist social relations and address questions of universality. We are immersed in a popular culture unswervingly saturated by endless spectacles meant to divert attention from substantive political issues and debates, and geared toward proselytizing in order to create silent accomplices in the ravages of corporate expansionism and imperialism. In the name of the most holy acts of consumption, the state media apparatuses, not only fail to resist the complete takeover of the public sphere by the logic of capital, but actively promote capitalist logic. In other words, under the guise of defanging the alienation produced by the social labor of capital, and making us more critically informed citizens, the media actively promote such alienation.

In order to address these issues and other related issues, critical pedagogy needs to be renewed—yes, it needs to bring itself face-to-face with the moment of the revolutionary. This time it has to be concerned with the problem of reasserting human action, and of finding forms of organization that facilitate human development. The depredations of progressive (i.e., left liberal) pedagogues have often subordinated praxis to the realm of ideas, theory, and the regime of the episteme. But critical socialist pedagogy recognizes the pivotal role of public political action, what has been called "public pedagogy." It's a pedagogy of revolutionary praxis. And here I would argue for a decolonizing, anticapitalist pedagogy. I have talked already about an anticapitalist pedagogy so let me explain what I mean by a decolonizing pedagogy. As I have written elsewhere, decolonizing pedagogical approach supports progressive initiatives such as smaller class sizes, improved low environmental impact school buildings, an end to school tracking, schools created on a human scale within or as local to communities as possible, cooperation between schools and local authorities rather than competition within the marketplace, vastly increased funding for education, increased powers for local governments to redistribute resources and participate in the development of antiracist, antisexist, and anti-homophobic policies and practices, and egalitarian policies designed to assist in more equal educational outcomes, irrespective of social class, gender, race, sexuality, or disability, and a curriculum geared toward socialist cooperation and ecological justice. But it also goes well beyond these initiatives. Decolonizing pedagogy in this instance does not only mean developing classroom strategies designed to contest neoliberal policies and practices, imperialism and militarism; it refers as well to developing a language of critique in which the concentration of corporate and state power is fundamentally challenged transnationally as well as locally. It is designed to understand society as a totality. Decolonizing educators realize that the concept of globalization alone is inadequate for understanding political and economic imperialism, wars of conquest and the pursuit of empire.

The decolonizing pedagogy that is being advocated here recognizes that as we exercise our neocolonial means of exploiting other countries (as the United States and other foreign capital have exploited the labor power of local populations, drawing them into the worldwide labor market), the mass media and culture in general constitute the central means by which the consent of the popular majorities are secured by the transnational capitalist class in pursuit of the consolidation of their profit-making practices. An important condition of possibility for economic exploitation is the subjective subordination of the popular majorities through education, entertainment, literature, and art. Such strategies of subordination are made more transparent within a decolonizing pedagogy that employs critical media literacy in the manner suggested by philosophers such as Douglas Kellner.

Decolonizing pedagogical practices are fundamentally activities rather than a contemplation of abstract concepts; they are designed to undermine empire by creating connections between the subjective feelings of alienation experienced by students and an understanding of their objective location within the social division of labor. In other words, the project of decolonization involves a concrete historical struggle and not a struggle for an abstract utopia. It involves providing students with opportunities for learning some of the basic quantitative and qualitative tools of urban sociologists and activists, for undertaking analyses and projects in their neighborhoods and communities, and within the schools themselves.

More fundamentally, decolonizing pedagogy is the creation of an historical identity through understanding the origins of the system that produces the alienation and estrangement experienced by students. In helping students analyze how the symptoms of their alienation are connected to the objective conditions of class society, teachers contribute to opening up a relationship between students and the historical present. The overall purpose is to undermine the established social relationship between classes, individuals and groups as well as the state's overdetermined systems of meaning such that it is possible to redefine what it means to be human outside of the repressive restrictions of the state. What is at stake here is more than following a methodology but developing the historical character of our social being. For instance, some radical educators such as Jeff Duncan Andrade and Ernest Morrell are teaching high school students to become radical sociologists that can analyze their own schools as institutions of domination, colonization, and social control. They call their approach, "thug life pedagogy" after the late hip-hop artist Tupac Shakur. Here critical pedagogy constitutes the building blocks for a relation with other people. In doing so, critical teaching helps hope resume its odyssey of struggle against the obstacles of fear, ignorance, and self-doubt. Tupac Shakur died at age twenty-five. His theory of humanization was called THUG LIFE (The Hate U Gave Little Infants Fucks Everyone). Tupac used to call youth fighting against oppression as the "roses that grow from concrete." According to Duncan-Andrade, "they are the ones who prove society's rule wrong by keeping the dream of a better society alive, growing despite the cold, uncaring, un-nurturing environment of the concrete." Andrade's students create block-umentaries where groups of students organized by neighborhood document how the historical, sociological, psychological, and educational tools of oppression are being used on their blocks to keep them and their families down. I think that's one way to utilize critical pedagogy. Of course, in doing all of this, it is also important to try to imagine what a postcapitalist project might look like on the ground, in the streets—how would it look at the level of the system and structure, the state apparatuses and the lifeworld. These are challenges that as educators we need to face.

S.L.: Contemporary reality is contaminated with the neoliberal capitalist ideology of values that from the perspective of the current world's system has no intention to change. How can we make on the one side this process more visible and understandable to others, and how can we decontaminate it on the other?

P.M.: Well, first of all, we have to recognize what you just reiterated in your question. That the current world system has no intention of changing. So what's the point of doing nuanced and rigorous analytic work, and sophisticated and hard-knuckled research on capitalist neoliberalism if it won't make a dent in the system? We need to stop being academics and start becoming activists. A year or so ago, I was placed on a list by a right-wing group, right at the top of the list known as the "dirty thirty" and denounced as UCLA's most dangerous professor—and the group that was responsible for this, a pro-Bush group, offered one hundred dollars to students who would secretly audiotape me, and fifty dollars for lecture notes they made in my classes (or classes of other leftist professors). This story became international because of its resemblance to McCarthyism and was characterized as a resurgent McCarthyism brought on by the assault on civil rights by the Bush cabal. This would have been much more unlikely prior to September 11, 2001. I think that the erosion of civil rights and the movement toward fascism here in the United States has really laid bare the serious threat we have to academic freedom here in the United States. We need to recognize that the social sciences themselves have helped to paralyze academics. There is only so much you can do when you are ensepulchred within a system of liberal modernity, liberal individualism, and the doctrine of neoliberal economics. As Takis Foutopolis points out, the negative conception of freedom that is embedded in liberal democracy—that is, as the absence of constraints—has been abstracted from its economic base, leading to an erosion of "formal freedoms." They are now being taken away in the semi-totalitarian regimes of the West, as the heteronomous society in which we live is being usurped by the elites who serve the transnational capitalist class. Education helps the elites in their attempts to control the popular majorities, let's face it. Not just education in classrooms, but a larger educational practice that takes place in the realm of communication. I am talking about the perpetual pedagogy of the corporate media. Clearly, this is what representative democracy thrives upon. So in terms laid out by Foutopolis, we need to struggle for an autonomous society in which the public space encompasses the entire citizen body, and where we can exercise a direct democracy (what he calls an inclusive democracy) where decisions at the macro-level (i.e., economic and political decisions) are part of an institutional framework of equal distribution of political and economic power among citizens. Here we have a different conception of freedom—not freedom from constraints, but freedom to achieve self-determination and participate in society's reflective and deliberative activities and bring a substantive content to the public sphere. The view of freedom as an absence of constraints (i.e., on the market, on how much one can own, on how much power a single corporation can accumulate, or on how much one can suffer need) has set the conditions of possibility for the capitalist class (or what we can now refer to as the transnational capitalist class) to defend its historical advantage, its existing hierarchies of power and privilege by out-maneuvering and manipulating the system so that it continually serves their own interests at the expense of the poor and the powerless. So how do we create this society of autonomous individuals? We need to move beyond the modern hierarchical society. As educators we can't take the existing social system for granted. And this means fighting in our work not only against the marketization of scientific research but also against both the objective rationalism of science (that enforces the neutrality of knowledge) as well as its postmodern relativism, as argued by Foutopolis. Teresa Ebert and others have warned us against pedagogical practices developed by the poststructuralist avant-garde who theorize experience in relation to trauma, desire, and affective relations in general as if these relations were antiseptically cleaved from relations of class, thereby replacing a conceptual analysis of the social totality with liberating pedagogical narratives grounded in local affective strategies—strategies that serve unwittingly as epistemological covers for economic conditions that help the subject cope with the objective material conditions of capitalist exploitation. So that most progressive pedagogical practices are little more than coping strategies that help students survive capital rather than transform it. This leads ultimately to a de-historicization of social life, and draws attention away from the way in which all human beings who populate capitalist societies are implicated in some manner in international class struggles and the social division of labor.

As the bloody fist of U.S. imperialism continues to hide itself inside a rose petal glove, it is important to see its support of the self-determination of Georgia and the Ukraine as part of a neo-imperialist practice of creating a neo-imperialist client state. Or what about Venezuela and Bolivia? Here the United States is trying to destabilize the government of Hugo Chavez and foster violence in the areas of Santa Cruz in Bolivia. Chavez is trying to free the poor from involuntary servitude to the forces of capital by, at the very least, fostering a vision of *un otro mundo*. In Bolivia we are witnessing opposition to Evo Morales by means of the violent tactics of the National Democratic Council (CONALDE), composed of five provincial governors, business associations, conservative civic groups, and legislators of the right-wing Podemos party led by former president Jorge Quiroga, who are working now with the UJC (Unión Juventud Cruceña, or Union of Santa Cruz Youth) and the Santa Cruz Civic Committee. Not to see these struggles as class struggles is to miss the main point.

S.L.: The production of passiveness is a strategy led by the neoliberal system that enables the system to produce social and political apathy. In this perspective how can we fight against the production of passiveness and how can we be (re)active to an already passive society so that it will be able to change the current neoliberal system?

P.M.: Well, first of all, we need to free ourselves from the *bête noir* of the progressives—positivism—which actually undergirds much of progressive educators' own work. One of the cardinal principles of positivism (that can be traced as far back as Hume and that involves a synthesis of idealism and empiricism) that was grandly pronounced as an antidote to the metaphysical belief in innate reason is that because all knowledge is derived through the senses, and because it is mediated by our subjectively conceived experiences through various conceptual system or systems of intelligibility, the objects of our contemplation can never really be known. In other words, our subjectively conceived experience mediates reality such that we can never know it objectively but only approach it through systems of mediation that form an insuperable barrier—a necessary wall of mystification. This has led to a passive theory of knowledge via a doctrine of experience that rejects understanding the world as a whole and resembles an empty solipsism where reality is reduced to a set of formal or logical statements (which leads to the existence of concepts because they are believed). This is a position that John Hoffman has called "positivism with a 'left' face." We can see this position reflected in the views of vulgar cultural relativists who believe that there is no real truth when it comes to values, and that one culture's values is no better than another culture's values and there is no basis for judging the value's of one culture over another. Values in this view are subjectively held commitments of certain groups of people and not objective truths—they are, in other words, merely what different cultures hold to be true (I will discuss the issues of cultural values a bit later on in my answer and how such vulgar relativism leads to the imposition of Western subjectivism that can lead to what Grosfoguel calls epistemic genocide or epistemicide). My position is Hegelian/ Marxist in the sense that I believe that we can't understand isolated bits of experience adequately without the whole—the absolute. We need to ask what makes experience possible, why do certain experiences count more than others, and what are the conditions of possibility for certain types of experiences. We read these dialectically against the absolute. But here I must make a caveat. Quijano warns us that when thinking about totality, or sociohistorical totality, we need to avoid the Eurocentric paradigm of totality. We can do this by thinking of totality as a field of social relations structured by the heterogeneous and discontinuous integration of diverse spheres of social existence, every one of which is in turn structured by its own historically heterogeneous, temporally discontinuous, and conflictive elements. Each element, however, has some relative autonomy, and can be considered a particularity and singularity. But they move within the general tendency of the whole. We can't think of totality as a closed structure. Change affects components in a historical field of social relations in a heterogeneous and discontinuous manner, and history does not move from one homogeneous and continuous whole to another. We need to avoid the perception of totality as seen from Europe, for instance, and yes, we can see that in terms such as "pre-capitalist" or "preindustrial" or "premodern." As Ramon Grosfoguel pointed out at a recent workshop I participated in, so-called primitive accumulation has always existed in Latin America but it took David Harvey to bring this term into the spotlight recently in his excellent work on accumulation by dispossession.`

I think it is important that we find a way to illuminate what is taken to be the natural (yet illusory) relation of human beings under capitalism and in doing so create a more "active" theory of knowledge. We can't have an active politics built on a passive theory of knowledge. That's one of the major issues. The current pedagogical concern with "experience" conceals from human beings that men and women are themselves the creators of these social facts and there are no supportable reasons why we should accept the naïve but perhaps historically inevitable illusion of the inviolability and necessary persistence of capitalism as the truth. This is why, for instance, I reject the position of Laclau and Mouffe, which denies that the material world has any significance outside of discursive articulation since I presuppose that there are real material interests that can and should be articulated. Of course, at the same time I reject the concept of totality as some aprioristic abstract schema imposed as a prefabricated mould upon reality. And, of course, I agree that Marxism may be misapplied in a militant or dogmatic manner, but it cannot be dismissed solely in non-sequitur fashion on the basis that it constitutes a universal theory. If it is driven by the spirit of self-assertion and totality such that it marginalizes, demonizes, and excludes cultural others, then this is deeply objectionable. If that is the Hegelian absolute that confronts us, then we need to get beyond it. And many trajectories of Marxism, such as the Marxist humanist tendency, have done so.

I don't believe that human agents are reduced in historical materialist analysis to a one-sided determinism in which history repeats itself with crystalline inevitability. That is, I don't believe that human beings are relegated to a passive role in which they are swept away in a swift current consisting of historical laws of motion, of nature-imposed necessity. I flatly reject this kind of mechanical materialism just as I reject, in contrast, a post-Marxist radical contingency and determinacy of the social. We need to develop an active materialism that does not end up in a series of false dualisms such as human beings/nature, or individuals/society. We need a pedagogy of what I have called "history-making," which is a revolutionizing practice that challenges capitalism's ability to invert our capacity for self-reflection so that we cannot understand capitalism's origins. As Teresa Ebert and others have argued, passive contemplation is not enough to alter those conditions in which human beings enter independently of their will. Rather, human beings must work to humanize those conditions and circumstances that shape them. We need more than abstract principles by which to strive for social justice, such as those that guide antiracist and antisexist curricula or eco-sensitive procedures for living in harmony with the planet. We must also challenge and transform those material circumstances seemingly beyond our control and this means developing a pedagogy of class struggle.

To reach for freedom is not an act of transcending reality but of actively reshaping it as Michael Lebowitz has argued. Similarly, truth is not an account of what is but what needs to happen. Here I am indebted to classical Marxists such as Teresa Ebert and Mas'ud Zavarzadeh for helping us to escape the debilitating world of subjectivism and voluntarism.

Now what do I mean by class struggle? Well, I believe that it is more that an economic struggle between the propertied and propertyless, but is a political struggle directed at the state (and here the hegemonic class is created through a system of alliances of class fractions that can best unify the power bloc). And winning the battle for democracy means much more than cultivating an ethical distaste for exploitation; it means actively working to end it. The state is not a neutral site, it is not an autonomous region that miraculously floats above the messy world of class antagonisms. Many progressive educators fail to realize this and in their refusal to move beyond a reclamation of the public sphere and an embracing of an anemic and abstract conception of democracy and freedom, they unwittingly reflect the leftist face of the capitalist class in which appearances are created and preserved while reality is eroded. Here the state is viewed as a site where mechanisms to win consent are pivotal, and where the process of legitimization constitutes a struggle by competing groups with various social, economic, and political interests. But semi-autonomous zones of engagement can be struggled for, where alternatives to capital can be seized and pushed forward, and this is the role of educator in the decolonizing classroom.

It is important to note that my concept of materialism here and my belief in a world outside of our thoughts and experiences in no way denies the "objectivity in parentheses" mentioned by Walter Mignolo since I reject a mutually exclusive transcendental ontology that is inhospitable to other ontologies or to observing our own acts of observation. My defense of a universality of social justice (and socialism) for all—creating the conditions of possibility for freedom from necessity for all—in no way rejects the pluriversity of knowledges. In fact, it affirms the legitimacy of knowledge disqualified by the imperiality of Eurocentered epistemology. But at the same time I am not prepared to give up on a Marxist humanism, on knowing the world through experiences that form a coherent view of both the self and social identity—with protagonistic actions in and on the world, a world in which the social is both the condition and outcome of human agency. And that also means I am not prepared to give up the fight for socialism. We struggle here for the simultaneity of universal and particular rights. And in doing so we don't privilege the idea of culture as a signifying system, but as a form of embeddedness in the materiality of social life. In doing so we can't abrogate the normative sense of what constitutes oppression, we consider it as a regulative idea. That's why I agree that we can advocate an epistemic cultural relativism (in the serious rather than the vulgar sense) in arguing that there is no privileged access to the truth, and that there is no direct correspondence between an object and its representation, but at the same time I am opposed to a judgmental relativism—that there are no grounds, rational grounds, for advocating some beliefs over others (we do this through dialoguing, in the spirit of respect, with other cultures, that is, respecting the nonsubjective character of their values). We can't fall into the trap of assuming all beliefs or arguments are equally valid. We need some explanatory adequacy or judgmental rationality in making decisions about various values, and we do this in relation to the material world, to things that exist independently of our attempts to explain them or account for them. And we don't impose this matrix of evaluation from our own Western geopolitical and epistemic location. And at the same time, following Aijaz Ahmad, we need new forms of politics that constitute human subjects both in their heterogeneity and in their universality. We can't abandon the challenge of universality, of universal rights, as a basis of solidarity and struggle. We need to struggle to make sure such universality is not representative of the bourgeois male, heterosexist, imperial, Christian colonizing subject, of course, but we can do this without abrogating the concept of universality. We can achieve this without making this an ontological philosophy of power that eclipses the trans-ontological and abrogates our radical responsibility to each other in the most debilitating forms of social amnesia (such as what we saw during the Bush administration in the United States). The principle of difference cannot provide us with the standards that oblige us to respect the difference of others, as Kenan Malik has pointed out. He notes that while difference can arise from equality, equality can never arise from difference. All universalisms are dirty, claims Bruce Robbins, and by this he means that universal standards are arrived at in conditions of unequal power. But he also notes that it is only dirty universalisms that will help us against the powers and agents of still dirtier ones. Similarly, we can't abandon every and all notion of essence, of something that is beyond appearance, because if we do that—if we privilege the concept of difference—then the notion of appearance alone becomes evidence that there are different categories of humanity that have little or nothing in common—or that they are incommensurable. If we do that, we fall into the same logic as positivist racial theory—which deduces categories of races from mere appearances of skin, hair, and bone. Apprehension of formal difference then moves to an explanation for the existence of different ontological categories. This is a dangerous move, just as I think a rejection of all humanism is dangerous, a rejection that follows the notion that modernity itself leads to an annihilation of the other. The barbarism of the twentieth century is, Kenan Malik argues, not so much a consequence of modernity as it was a product of specific capitalist social relations. I object to the dismissal of universalism on the grounds of its hostility to otherness and for its attempts to impose Eurocentric or Euro-American ideas of rationality on other peoples. I do believe it is important to critique the false claim of universality inherent in the European particular, of course. However, I agree with Eagleton when he argues that the Western, "First World," postmodern intelligentsia has mistaken "its own very local difficulties for a universal human condition in exactly the manner of the universalist ideologies it denounces." Those who are preoccupied with the "crisis of humanism" need to remember that it is not everyone's crisis (not even in the West). We need to remember, as well, that many Third-World struggles of the postwar era drew upon the "logic of universalism" and that it remains crucial in many struggles for liberation.

Some might just dismiss this as Eurocentric third worldism but such struggles could indeed set the conditions of possibility for decolonial struggles. Let's face it, humanist principles can be coercive or liberating, depending on who is employing them and for what purpose.

Here I am calling for a "nonabstract and nonhomogeneous" form of universalism as a *political* referent. We need to distinguish between an abstract universalism that dissolves important differences among diverse phenomena and a concrete universalism that carefully draws such distinctions while upholding conditions that are binding for all. I have written elsewhere that the restricted and often dangerously destructive Western bourgeois character of Enlightenment universalism is a worthy and necessary object of critique, but to attack the idea of universalism itself is problematic. We need to be wary of conflating universalism with uniformity, because universals can be both various and locally diverse. There is no question that, for instance, colonialism has been intrinsic to the kind of universality that we have had in much of world history and that the only universal civilization that exists today is global capitalism. The solution isn't getting rid of the concept of universalism but working toward a better universality—and some of this can be achieved in struggling for what has been called by Walter Mignolo and Grosdfoguel and others as pluriversality.

A more useful alternative to "dismissing universalisms as masked particularisms" is to side with Terry Eagleton (1996) who writes that to be a socialist is, among other things, to recognize that

> universality *doesn't* exist at present in any positive, as opposed to merely descriptive or ideological, sense. Not everyone, as yet, enjoys freedom, happiness, and justice. Part of what prevents this from coming about is precisely the false universalism that holds that it can be achieved by extending the values and liberties of a particular sector of humankind, roughly speaking Western man, to the entire globe. Socialism is a critique of this false universalism, not in the name of cultural particularism but in the name of right of everyone to negotiate their own differences in terms of everyone else's. (Emphasis in the original; p. 118)

But I need to flesh out further what I mean. Where do other cultures come into play? I am committed to the objective truth of the values of other cultures; I do not reject the nonsubjective character of their values at all. I am a cultural relativist not in the vulgar sense of believing at the philosophical metacultural level that cultural values are simply subjectively held commitments. Rather, I give serious weight to the objective truth of other cultures in the sense of maintaining that all cultures have their own access to truth even though the values of other cultures might not be compatible with the values and conceptual schemes of my own culture. One has to guard against the imposition of a Western subjectivism calcified into an imperial universal. As Nelson Maldonado-Torres so brilliantly argues, capitalism and coloniality have betrayed the damned at the expense of the trans-ontological, that is, at the expense of validating the knowledges and forms of being and the very humanity of the colonized, creating a conditions of life founded on receptive generosity, or at the expense of "altericity" (a term that Maldonado-Torres uses to capture the priority of the relationship of responsibility between self and others). The task then becomes, in Maldonado-Torres' view, eliminating what he calls "the coloniality of being" (the normalization of everyday warfare against colonial subjects that can be found in national ontologies and identitarian ontologies, etc.). He calls for the elimination of sub-ontological difference (ontological colonial difference or the naturalization of sub-Others as legitimate recipients of excessive violence) and restoring the meaning and relevance of trans-ontological difference (difference produced as a formative event in the production of being that makes possible communication between a self and an Other and the foundation of justice through the vertical relation between subjectivity and alterity).

As a Marxist humanist who finds a great deal of value in the writings of Hegel, let me take Hegel as an example of a philosopher of whom many multiculturalists are critical because of his undeniable racism and ethnocentrism. Of course, we must reject Hegel's ethnocentrism and racism. But Hegel's own philosophical position provides us with the tools to subvert Hegel's own ethnocentrism and cultural racism. Philip Kain's book *Hegel and the Other* makes a good case that Hegel takes the position that every culture stands before the absolute such that the absolute is the expression of that culture and that culture is the expression of the absolute. Paraphrasing Kain, Hegel believes that it is philosophy's task to construct the absolute for consciousness, and because we construct it doesn't mean the absolute isn't real. Knowing, for Hegel, is part of the absolute—so it is impossible to know anything, really, before we know—so that we need, in other words, to start without any epistemological criterion because (as Kain, Norman, and others remind us) to attempt to demonstrate an epistemological criterion that would claim to tell us what we can and cannot know already is a form of knowing and this would presuppose the criterion it was supposed to demonstrate. So, we begin with the absolute, that is, we begin with something unproven and by abstracting from the absolute we eventually realize that this is impossible so Hegel takes us full circle back to the absolute again by showing us how everything is internally related, how things are mutually constituted, that is, they are constituted by other things and fused by the whole, and this is what Marx's theory of internal relations was all about—and more. Each culture constructs the absolute for consciousness in a different way; in doing so, each culture has access to its own truth. So that this absolute needs to be seen as open. The human spirit can be realized only through the spirit of particular cultures and therefore we should not view the absolute as closed and we should welcome engaging cultural difference. In fact, the absolute demands its own subversion. We can't let the absolute be a totalizing juggernaut linked to Western imperial adventures, so we need to incite difference within the absolute. The absolute that is true for itself for a particular culture needs to be true in and for itself. It can't not be absolute in itself and be absolute for a particular culture. Thus, I am a cultural relativist in the manner that Kain describes, rather than a vulgar relativist. I am a cultural relativist who denies that cultural relativism is a self-refuting term insofar as I maintain that consciousness develops within a specific cultural context and a specific historical era or juncture. That being said, it is also the case that culture can embody or embed truth. I reject the existence of a space station platform above culture—some kind of Death Star standpoint where the emperor shrouded in a blood red cowl of Western epistemology stands above culture and exercises a supercultural God's eye view, some kind of sky hook or observation platform nested in the metaphysical heights of Mount Olympus or neutral common coordinate system outside of the Matrix from which to transvalue all the meanings generated by the universe and pronounce indubitable judgment on all convention. It's unlikely, at any rate, that different cultures are completely incommensurable since different systems of intelligibility or different conceptual systems can provide some translation from one culture to another. Like Kalin, I doubt very much that all truth can be captured within one system of intelligibility—at least consistently captured. We need to respect other conceptual systems—other perspectives—because no single system can capture all truths. I have not given up on translability. Paulo Freire once told me during a conversation we had about translating high theory into classroom practice that it was important to translate his work within the contextual specificity of where I was standing—where I was located—as a teacher and where my students were located as students. He told me that my job was one of translation—of translating his philosophical language into a language that impacted teachers in various pedagogical settings so that they could provide spaces of opportunity where critical learning could take place. That I should not simply impose his conception of knowing on my students—which would be a form of conceptual/cultural imperialism. Our misunderstandings with individuals from other cultures does not mean that these cultures are incommensurable. After all, our conceptual systems can be different—and we can find such systems to be incompatible or inconsistent or both—but there is always the possibility of translation, no matter how limited. I am not advancing a moral obligation for tolerating other cultures; I want to be clear on that. We need other grounds for taking a position against intolerance. We need to engage other cultures in order to better understand ourselves and others, our interdependence, and mutual obligations toward each other, and the world as a totality. We engage in the values and belief systems of other cultures, not in a vulgar way by according them only subjective merit and by denying them any nonsubjective validity. We engage them by embracing a concept of truth. As Kain points out, if we believe that all values are subjectively held commitments and do not constitute objective truths—then we are going to fall into the trap of vulgar relativism—and this leads to epistemic violence inevitably. We need a commitment to the objective truth of other cultural values. All cultures have access—their own access—to the truth. The rejection of the nonsubjective character of the values of other cultural beings (i.e., considering their views as just another position among many positions without any real truth value) is tantamount to the imposition of Western subjectivism. Hegel, of course, is ethnocentric but Kain argues persuasively that his system gives us a chance to subvert his ethnocentrism. Hegel was also a racist, and while Kain argues that he was not a theoretical or scientific racist, he certainly must be condemned for his ranking different ethnic groups according to their participation in the absolute and his view of European superiority. It is simply indefensible. Hegel's ignorance of Africa is staggering and his view of indigenous Americans must be criticized and denounced (see his ranking of cultures in *Philosophy of Mind*) as he not only describes Western ethnocentrism, imperialism, and racism but actually endorses them. Kain writes that in his rankings of races as higher and lower Hegel rejects a scientific justification of his ranking. Racial essences, which reside for Hegel in mind or spirit, are educable and are not unchanging. This, and the fact that Hegel includes different races for being part of the construction of the absolute does not mean we should forgive him for his racism and ethnocentrism but it does mean that we can look to Hegel's system for a way of subverting Hegel. Kain suggests that we need to view Hegel's exclusion of other cultures from a dialectical perspective—they all contain the truth—in other words, world spirit does not belong to any one nation. There remains an outside to Western epistemology in Hegel's overall philosophical system. In this way Hegel invites a subversion of his own narrative of cultural ranking since, according to Kain, Hegel is committed to heterogeneity and does not focus on racial purity. It is important to keep in mind that Hegel does not want a single world culture—he is a pluralist at heart. He does not seek to absorb other cultures into an abstract universal. But undeniably, Hegel is ethnocentric at the philosophical, metacultural level. But his system carries within itself a mechanism for subverting itself.

S.L.: Anibal Quijano in his text *Coloniality of Power, Eurocentrism and Latin America* wrote that historically only through the colonization of the Americas the capital could consolidate and obtain global predominance, establishing the new world order known as capitalism. Isn't it interesting and amazing that after five hundred years of modern world system the new socialist revolution is actually happening precisely in Latin America that is probably one of the most exploited geographical areas where colonialism through centuries allowed the expansion of capitalism (globally) up to now? Let me follow this with another question. It is quite obvious that if we want to build another perspective (and not only the capitalist one), we will have to engage in a fight against the current structure of power on all social levels. The fight against this hegemony is already going on but it's seems that it is still a little bit dispersed (structured in different theoretical fields). Does radical critical pedagogy work together with other important de-linking projects, such as decoloniality (W. Mignolo), and do you think that it is time for a historical revolution on a global scale to take place?

P.M.: I very much admire Quijano's work and that of Walter Mignolo. They need to be read in my field of education and of course in other fields as well. And of course there is the very important and I would say urgently needed work of Enrique Dussel, and Ramon Grosfoguel and Nelson Maldonado-Torres. But you mentioned Quijano. Well, for instance, Quijano draws attention to the production of a world system of classification, nonexistent before the sixteenth century, and deftly argues that previous forms of domination (e.g., gender) were then reconfigured around the new system of racial classification. Knowledge and society were organized around the category of race (a precursor to Eurocentrism) to the extent that the division of labor itself was naturalized. With the help of capitalism, the idea of race helped to yolk the world's population into a hierarchical order of superior and inferior people and it became a central construct in creating and reproducing the international division of labor, including the global system of patriarchy. Quijano is correct when he writes that "Domination is the requisite for exploitation, and race is the most effective instrument for domination that, associated with exploitation, serves as the universal classifier in the current global model of power." He also makes an important intervention when he argues that dualism and evolutionism situated the European subject as the most spiritually evolved while women and slaves were viewed as the most primitive, locked into their corporeality. Slavery, serfdom, wage labor, and reciprocity, all functioned to produce commodities for the world market. What Quijano calls a "colonial power matrix" ("patrón de poder colonial") affecting all dimensions of social existence such as sexuality, authority, subjectivity, and labor, Berkeley professor Ramon Grosfoguel conceptualizes as a historical-structural heterogeneous totality that by the late nineteenth century came to cover the whole planet.

Expanding on Quijano's work, Grosfoguel describes the coloniality of power as an entanglement of multiple and heterogeneous global hierarchies ("heterarchies") of sexual, political, epistemic, economic, spiritual, linguistic, and racial forms of domination and exploitation where the racial/ethnic hierarchy of the European/non-European divide transversally reconfigures all of the other global power structures. As race and racism became the organizing principle that structured all of the multiple hierarchies of the world-system, Grosfoguel argues that the different forms of labor that were articulated to capitalist accumulation on a world-scale were assigned according to this racial hierarchy. Cheap, coercive labor was carried out by non-European people in the periphery and "free wage labor" was exercised in the core. Such has been the case up to the present day. Grosfoguel (in press) points out that, contrary to the Eurocentric perspective, race, gender, sexuality, spirituality, and epistemology are not additive elements to the economic and political structures of the capitalist world-system, but a constitutive part of the broad entangled "package" called the European modern/colonial capitalist/patriarchal world-system. This work is very important for educators to engage. We need to enter into dialogue with this line of work—I agree. But in some of the work based on moving beyond Eurocentrism there is a critique of Marxism that I think, while it does a brilliant job of critiquing the Marxism of the Internationals, what could be called militant manifesto Marxism, it also ignores the Marxist humanist standpoint and some work in classical Marxism by scholars such as Teresa Ebert and Mas'ud Zavarzadeh, and educationalists such as Valerie Scatamburlo-D'Annibale, Glenn Rikowski, Mike Cole, Deb Kelsh, Dave Hill, Paula Allman, and others. The last several decades have witnessed a concerted attack on "totality" by progressive educators of various stripe who purport to challenge the totalizing grand narratives informing Western domination and Eurocentrism. Decentering and recentering representational narratives has become the order of the day. Critiques of political economy and capitalist relations of exploitation have given way to a decentering of systems of signification that function within apparatuses of power-knowledge. The dualism of oppressor/oppressed has been replaced with the ambivalence of cultural difference and the notion that every subject is "always already oppressor and oppressed." The cultural differences of "in-between-spaces" are prioritized over the totalizing discourses of universal liberation. And while I recognize, with Gramsci, that consciousness can be "contradictory" I do not see the struggle for social justice as exploiting the ambivalence inherent in power relations or as a rupture of a signifying chain, but as the revolutionary praxis of the oppressed forged in class struggle. Some of the work on coloniality and identity, in their theorizations of "difference" (usually in their discussions of race) circumvent and undermine any systematic knowledge of the material dimensions of difference and tend to segregate questions of difference from class formation and capitalist social relations. This is not the case with Quijano and Grosfoguel and Mignolo, which is why their work is so important for educators to engage. But some of the postcolonial scholars (those who work to consolidate "identitarian" understandings of difference based exclusively on questions of cultural or racial hegemony) tend to downplay or basically ignore the totality of capitalist social relations. Therefore, I think it is important, as I have argued with Valerie Scatamburlo-D'Annibale, to (re)conceptualize difference by drawing upon Marx's materialist and historical formulations. Difference needs to be understood as the product of social contradictions and in relation to political and economic organization. We need to acknowledge that "otherness" and/or difference is not something that passively happens, but, rather, is actively produced. Drawing upon the Marxist concept of *mediation* can help us to unsettle our categorical approaches to both class and difference, for it was Marx himself who warned against creating false dichotomies in the situation of our politics. He argued that it was absurd to "choose between consciousness and the world, subjectivity and social organization, personal or collective will and historical orstructural determination." In a similar vein, it is equally absurd to see difference as a historical form of consciousness unconnected to class struggle. So we need very much to examine the institutional and structural aspects of difference, the meanings that get attached to categories of difference, and to understand how differences are produced out of, and lived within specific historical formations.

While it is important to create greater cultural space for the formerly excluded to have their voices heard (represented), at the same time we have to make sure that this does not simply reinscribe a neoliberal pluralist stance rooted in the ideology of free-market capitalism. In short, cultural politics becomes, in this case, modeled on the marketplace and freedom amounts to the liberty of all vendors to display their "different" cultural goods. What many postcolonial theorists fail to address is that the forces of diversity and difference are allowed to flourish provided that they remain within the prevailing forms of capitalist social arrangements. The neopluralism of difference politics (including those based on "race") cannot adequately pose a substantive challenge to the productive system of capitalism that is able to accommodate a vast pluralism of ideas and cultural practices, and cannot capture the ways in which various manifestations of oppression are intimately connected to the central dynamics of capitalist exploitation. This is where Marxist humanist analysis can help.

Why am I saying all this? Because the struggle for diversity through antidiscrimination has actually seen U.S. society adapt to its production of inequality and make the politics of difference work to bolster neoliberal capitalism. As you know, I have been creating antiracist, antisexist, and anti-homophobic curricula and pedagogical practices for decades, and have been challenging educational policies and practices from these same perspectives, but I have also been stressing the strategic centrality of class struggle. It is not that I am trying to privilege class over race and gender or reduce race to an artifact of class, as some of my critics claim. Rather, I am trying to make the case that unless antiracist and anti-patriarchal struggles are multipronged endeavors and conjugated with class struggle, their efforts could lead to a strengthening of inequality rather than bringing about its defeat. Why? Because while capitalist society is becoming less discriminatory, it is increasingly becoming unequal, not more equal. In a recent article in the *New Left Review* (2008, p. 33; not a journal I read very much these days), Walter Benn Michael has put the case succinctly and powerfully in the following terms:

> In 1947—seven years before the Supreme Court decision in Brown v. Board of Education, sixteen years before the publication of Betty Friedan's *The Feminine Mystique*—the top fifth of American wage-earners made 43 per cent of the money earned in the us. Today that same quintile gets 50.5 per cent. In 1947, the bottom fifth of wage-earners got 5 per cent of total income; today it gets 3.4 per cent. After half a century of anti-racism and feminism, the us today is a less equal

society than was the racist, sexist society of Jim Crow. Furthermore, virtually all the growth in inequality has taken place since the passage of the Civil Rights Act of 1965—which means not only that the successes of the struggle against discrimination have failed to alleviate inequality, but that they have been compatible with a radical expansion of it. Indeed, they have helped to enable the increasing gulf between rich and poor. Why? Because it is exploitation, not discrimination, that is the primary producer of inequality today. It is neoliberalism, not racism or sexism (or homophobia or ageism) that creates the inequalities that matter most in American society; racism and sexism are just sorting devices. In fact, one of the great discoveries of neoliberalism is that they are not very efficient sorting devices, economically speaking.

Michael claims that the debates about race and gender are essentially empty unless the focus remains on capitalist exploitation. He goes on to say the following:

Whether debates about race and gender in American politics involve self-congratulation, for all the progress the us has made, or self-flagellation over the journey still to go, or for that matter arguing over whether racism or sexism is worse, the main point is that the debate itself is essentially empty. Of course discrimination is wrong: no one in mainstream American politics today will defend it, and no neoliberal who understands the entailments of neoliberalism will do so either. But it is not discrimination that has produced the almost unprecedented levels of inequality Americans face today; it is capitalism.

In my mind, Michael clearly identifies the central flaw of much left politics today, whether in education or any other field. Given the current liberal pluralist framework that separates out economic rights from human rights, the most that can come of a multiculturalism—or critical pedagogy for that matter—is one that is guided by a neoliberalism of the right or a neoliberalism of the left. Here is a section of his warning:

[I]t is clear that the characterization of the race–gender debate as "empty" needs to be qualified. For the answer to the question, "Why do American liberals carry on about racism and sexism when they should be carrying on about capitalism?," is pretty obvious: they carry on about racism and sexism in order to avoid doing so about capitalism. Either because they genuinely do think that inequality is fine as long as it is not a function of discrimination (in which case, they are neoliberals of the right). Or because they think that fighting against racial and sexual inequality is at least a step in the direction of real equality (in which case, they are neoliberals of the left). Given these options, perhaps the neoliberals of the right are in a stronger position—the economic history of the last thirty years suggests that diversified elites do even better than undiversified ones. But of course, these are not the only possible choices.

So why I am so hard on the poststructuralists and many of the postcolonial theorists? For precisely this reason. By not conjugating anti-racism and anti-sexism with class struggle, efforts at ending discrimination could help to foster inequality. Consequently, critical pedagogy needs to remove itself from a left-liberal politics in which its antipathy for challenging neoliberal capitalism only strengthens the unholy grip of capital on the poor and the powerless. This is essentially the message of what I have been calling revolutionary critical pedagogy. Now it is quite possible for Michael to make a mistake here, and that would be to concentrate only on capitalism, and forget about colonial epistemology and pluriversality and the fact that multiple antagonisms are co-constitutive. If we want to challenge capitalism, we need to delink from the Western rational episteme in the sense that such a de-linking helps us recognize that it is the desire for wealth and capital accumulation that helps to provide the epistemic oxygen that nourishes capital. In other words, we need to negate capital but we need to negate what we have negated, and in doing so we can learn a lot from Hegel if we take seriously his concept of self-referential negation and what Raya Dunyevskaya referred to as absolute negativity. And we can learn a lot from epistemologies outside of Western ones, the epistemologies of the *damnés*, or what Fanon referred to as the racialized and colonized peoples of what some call the Third World—those who inhabit locations in the colonial matrix of power that are much different than our own, for instance. I am thinking here of the importance of the ethico-political project of the Zapatistas, whose geopolitics and body-politics provide the substance for what Mignolo calls "border thinking." But at the same time I don't want to downplay the importance of some Western philosophical traditions—such as critical theory and Marxist theory—that can only be strengthened by our efforts to decolonize them; we need also to recognize their power and promise for creating a praxis of liberation that can challenge the power complex that Grosfoguel calls the "European/Euro-North American capitalist/patriarchal odern/colonial world-system." The Marxist humanist tradition, for instance, cannot be simply dismissed as one totalizing episteme seeking to appropriate all of subjectivity into its voracious vortex of Western epistemology.

S.L.: You are working a lot with countries in Latin America (Bolivia, Venezuela) that are leading the socialist project (socialist revolution) of the twenty-first century. Slovenia as one of the states of the former Federal Republic of Yugoslavia had already been part of a similar socialist project with socialist self-government and collective (i.e., public) property. Soon after the fall of the Berlin Wall we entered the neoliberal order. Neoliberal ideology took place when the newborn Slovenian capitalists allowed for a reorganization of socialism into neoliberal capitalism. All socialist values have been replaced by the neoliberal ones. The same story happened in all the countries of the former Eastern European socialist bloc. Therefore how can we imagine a socialist alternative to neoliberalism and what are the real possibilities for establishing a socialist democracy in countries such as Bolivia and Venezuela or elsewhere in the world?

P.M.: Yes, it has been a tragic history. The whole dismantling of the socialist project by the West. I hope you will permit me to answer this question philosophically. But philosophy here means changing the world, not just interpreting it. But how to envision a new beginning? That is the challenge of our times. I believe we need to focus on human development through a renewed understanding of what the negation of the negation means. Because it will take us to a place of absolute negativity, and that is where we can forge spaces of hope and possibility. Here I return to my Marxist humanist roots and to the work of Raya Dunayevskaya and Peter Hudis. Hudis (2005) notes that the genius of Hegel was that he was fully aware that negation is dependent on the object of its critique. In other words, ideas of liberation are impacted, in one way or another, by the oppressive forms that one tries to reject, and that negation per se does not totally free one from the negated object. But unlike the postmodernists that centuries later followed him, Hegel believed that there was a way for negation to transcend the object of its critique. He therefore introduced the notion of "the negation of the negation." Hudis makes clear that the negation of the negation, or second negativity, does not refer simply to a continuous series of negations—that can potentially go on forever and still never free negation from the object of its critique. Hegel instead argues for a self-referential negation. By negating itself, negation establishes a relation with itself—and therefore frees itself from dependence on the external object. According to Hudis, this kind of negativity, second negativity is "absolute," insofar as it exists without relation to another outside itself. In other words, negation is no longer dependent on an external object; it negates such dependency through a self-referential act of negation.

According to Hudis, Marx did not dismiss the concept of the "negation of the negation" as an idealist illusion but instead appropriated the concept of the self-referential negation "to explain the path to a new society." Marx understood that simply to negate something still leaves us dependent upon the object of critique; in other words, it merely affirms the alienated object of our critique on a different level. As Hudis (2000) states, and Dunayevskaya and other Marxist humanists have pointed out, that has been the problem with revolutions of the past—they remained dependent upon the object of their negation. The negation of the negation, however, creates the conditions for something truly positive to emerge in that absolute negativity is no longer dependent on the other. Here is how Hudis (2005) puts it in his own words, using the example of communism:

> Communism, the abolition of private property, is the negation of capitalism. But this negation, Marx tells us, is dependent on the object of its critique insofar as it replaces private property with collective property. Communism is not free from the alienated notion that ownership or *having* is the most important part of being human; it simply affirms it on a different level. Of course, Marx thinks that it is necessary to negate private property. But this negation, he insists, must itself be negated. Only then can the *truly positive*—a totally new society—emerge. (Emphasis in the original)

Marx believed that labor or human praxis can achieve the transcendence of alienation but what was needed was a subjective praxis connected with a philosophy of liberation that is able to illuminate the content of a postcapitalist society and project a path to a totally new society by convincing humanity that it is possible to resolve the contradiction between alienation and freedom. We can't resolve such a contradiction within the social universe of capital and capital's value form of labor. In Hudis' (2005) terms, we need to concretize "Absolute Negativity *as New Beginning,* rather than repeating the truths of an earlier era that no longer have the power to seize humanity's imagination." And talking about seizing the imagination, well, that is certainly happening in Latin America with Bolivia, Venezuela, and other places.

So what am I saying here? In conjunction with the centrality of class struggle, we need today a new form of philosophical comprehension and working out of what absolute negativity as a new beginning means today. We need to start to define the characteristics of a world outside of capital's value form—and we need the concepts of plurversality and transmodernism to help us in doing so. We need, in other words, not only to acknowledge the priority of material necessity in historical development but to start constructing a postcapitalist society, what we refer to as a socialist society in dialogue with other cultures, other values. This understanding of absolute negativity as a seedbed for new beginnings is the motor of a renewed critical/revolutionary pedagogy guided by the imperative of class struggle, and the development of a decolonial philosophy of praxis. Now to a certain extent this is happening in Venezuela, where I have been invited to develop radical pedagogical alternatives. It is not a socialist society but the Chavez government is trying to create the conditions of possibility for socialism to emerge. And great experiments are taking place in attempting to create spaces for human development. The teachings of Che Guevara have taught me a lot about the importance of human development. So many revolutionary leaders of his day thought that when the iron laws of capital were smashed by nationalizing the industries and centralizing the economy that socialist society would automatically spring forth. Che knew otherwise. He knew that socialist society needed socialist development—a socialist human development and the creation of new men and women.

During a visit to the University of Las Villas, where he received the *honoris causa* in the School of Pedagogy in the presence of thousands of students, he delivered his well-known speech on the role of intellectual activity in bringing about social equality. On this important occasion he called for that university—as well as other universities around the country—to identify themselves fully with Cuba's ordinary people and workforce; to become cognizant of "social outsiders," particularly those oppressed by racism and economic inequality. Fernández Retamar (1989) has a wonderful quotation from Guevara, who sought to propose to the university professors and students the kind of transformation that all of them would have to undergo in order to be considered truly useful to the construction of a socialist society:

> I would never think of demanding that the distinguished professors or the students presently associated with the University of Las Villas perform the miracle of admitting to the university the masses of workers and peasants. The road here is long; it is a process all of you have lived through, one entailing many years of preparatory study. What I do ask, based on my own limited experience as a revolutionary and rebel commandante, is that the present students of the University of Las Villas understand that study is the patrimony of no one and that the place of study where you carry out your work is the patrimony of no one—it belongs to all the people of Cuba, and it must be extended to the people or the people will seize it. And I hope—because I began the whole

> series of ups and downs in my career as a university student, as a member of the middle class, as a doctor with middle-class perspectives and the same youthful aspirations that you must have, and because I am convinced of the overwhelming necessity of the revolution and the infinite justice of the people's cause—I would hope for those reasons that you, today proprietors of the university, will extend it to the people.

Well, we can only hope along with Che that the people will become the proprietors of our universities. But, as Che knew only too well, hope is not enough. We need to bring hope into the activist orbit of decolonial struggle. And that is our purpose as critical educators.

Bibliography

Dussel, Enrique. (1985). *Philosophy of Liberation*. New York: Orbis.

Eagleton, Terry. (1996). *The Illusions of Postmodernism*. New York, NY: Wiley-Blackwell, p.118

Ebert, Teresa. (In press). *Cultural Critique (With an Attitude)*. Champaign, Illinois. University of Illinois Press.

Ebert, Teresa, and Zavarzadeh, Mas'ud. (2008). *Class in Culture*. Boulder and London: Paradigm Press.

Grosfoguel, Ramon. (2005). "The implications of subaltern epistemologies for global capitalism: Transmodernity, border thinking, and global coloniality." In R.P. Applebaum and W.I. Robinson (eds.). *Critical Globalization Studies* (pp. 283–292). London: Routledge.

Gržinić, Marina. (2008). De-linking epistemology from capital and pluriversality—a conversation with Walter Mignolo, Part 1. *Reartikulacija*, no. 4, Summer. Retrieved from http://www.reartikulacija.org/dekolonizacija/dekolonialnost4_ENG.html.

Hegel, Georg Wilhelm Friedrich and Wallace, William. (1971). *Hegel Philospohy of Mind*. USA: Oxford University Press.

Hoffman, John. (1975). *Marxism and the Theory of Praxis*. London: Lawrence and Wishart.

Hudis, P. (2000). Can capital be controlled? *News & Letters Online*, at http://www.newsandletters.org/Issues/2000/April/4.00_essay.htm

———. (2005). "Marx's critical appropriation and transcendence of Hegel's theory of alienation." A paper delivered to a session on "Alternatives to Capitalism" in the Left Forum in New York City, March 2005.

———. (2008). Hegel's phenomenology today: a Marxist-humanist view. *News & Letters*, December, 2007–January, 2008.

Kain, Philip J. (2005). *Hegel and the Other: A Study of the Phenomenology of Spirit*. New York: State University of New York Press.

Malik, Kenan. (1997). "The mirror of race: postmodernism and the celebration of difference." In E.M. Wood and J.B. Foster (eds.). *In Defense of History: Marxism and the Postmodern Agenda* (pp. 112–133). New York: Monthly Review Press.

Maldonado-Torres, Nelson. (2008). *Against War: Views from the Underside of Modernity*. Durham, North Carolina: Duke University Press.

Michaels, Walter Benn. (2008). "Against diversity." *New Left Review*, 52, July–August. As retrieved from: http://newleftreview.org/?page=article &view=2731.

Mignolo Walter. (1987). *The Darker Side of the Renaissance: Literacy, Territoriality and Colonization*. Ann Arbor: The University of Michigan Press, 1995.

———. (1999). *Local Histories/Global Designs: Coloniality, Subaltern Knowledges and Border Thinking*. Princeton: Princeton University Press.

———. (2005). *The Idea of Latin America*. London: Blackwell.

Quijano, Anibal. (1993). "'Raza,' 'Etnia' y 'Nación' en Mariátegui: Cuestiones Abiertas" [Race, ethnicity and nation in Mariategui: Open questions]. In R. Forgues (ed.). *José Carlos Mariátguiy Europa: El Otro Aspecto del Descubrimiento* (pp. 167–187). Lima, Perú: Empresa Editora Amauta S.A.

———. (1998). "La colonialidad del poder y la experiencia cultural latino-americana" [The coloniality of power and the Latin America cultural experience]. In R. Briceño-León and H.R. Sonntag (eds.), *Pueblo, época y desarrollo: la sociología de América Latina* (pp. 139–155). Caracas: Nueva Sociedad.

Quijano, Anibal and Ennis, Michael. (2000). Coloniality of power, ethno-centrism, and Latin America. *NEPANTLA, 1,* 533–580.

Quijano, A., and Wallerstein, I. (1992). "Americanity as a concept, or the Americas in the modern worldsystem." *International Journal of Social Sciences,* 134, 583–591.

Retamar, Roberto Fernández. (1989). *Caliban and Other Essays.* Minneapolis, MN: University of Minnesota Press.

Part II

Strategies for Practicing the Pedagogy of Critique

Chapter 5

Class, Capital, and Education in this Neoliberal and Neoconservative Period

Dave Hill

Introduction

In the aftermath of "the Bankers' Bailout," the 2008–2009 crisis of Neoliberal Finance Capital (and its subsequent impacts on Industrial Capital), the political response to "the credit crunch," by parties funded by Capital, such as the Democrats and Republicans in the United States and Labour, Liberal, and Conservative Parties in the United Kingdom, and conservative and social democrat parties globally, is not to blame the capitalist system. Not even to blame the neoliberal form of capitalism. Instead they promise—indeed, threaten more severe forms of neoliberalism—and control. The current neoliberal project, the latest stage of the capitalist project, is to reshape the public's understanding of the purposes of public institutions and apparatuses, such as schools, universities, and libraries. In schools, intensive testing of pre-designed curricula (high-stakes testing) and accountability schemes (such as the "failing schools" and regular inspection regime that somehow only penalizes working class schools) are aimed at restoring schools (and further education and universities) to what dominant elites—the capitalist class—perceive to be the schools' "traditional role" of producing passive worker/citizens with just enough skills to render themselves useful to the demands of capital.

In the United States and the United Kingdom and throughout other parts of the globe (Hill, 2005b, 2006a, 2009; Hill and Kumar, 2009; Hill et al., 2006), policy developments such as the 1988 Education Reform Act, passed by the Conservatives and extended/

deepened by New Labour, and in the United States, the Bush "No Child Left Behind Act" of 2001 have nationalized and intensified patterns of control, conformity, and (increasing) hierarchy. These and other policies such as the Patriot Act in the United States, which among other things permits secret services to spy on/access the library borrowing habits of readers, have deepened the logic and extent of neoliberal capital's hold over education reforms, over public services. They are an attempt to both intimidate and to conform critical and alternative thinking.

In the United States, such reforms include the heavy involvement of educational management organizations (EMOs) as well as the introduction of voucher plans, charter schools, and other manifestations of the drive toward the effective privatization of public education. England and Wales, meanwhile, have endured the effective elimination of much comprehensive (all-intake, all-ability) public secondary schooling. Commercialization and marketization have led to school-based budgetary control, a "market" in creating new types of state schooling, and the effective "selling off" of state schools to rich and/or religious individuals or groups via the Academies scheme. The influence of neoliberal ideology also led to the October 2005 proposals for state schools, which have historically fallen under the purview of democratically elected local school districts, to become independent "mini-businesses" called "independent trust schools" (Hill, 2006). Similar attempts at change have occurred throughout developed and developing countries (Hill, 2005a, 2009; Hill et al., 2006, Hill and Kumar, 2009; Hill and Rosskam, 2009).

However, the impact of the "New Labour" government in Britain on society and our schools and universities and the impacts of the Bush administration in the United States make it impossible to understand the current crises in schools and in democracy solely in terms of neoliberalism. We need also to consider the impact of neoconservatism.

In this chapter, I want to provide an overview of how those agendas in education play themselves out in the United Kingdom, the United States, and worldwide.

Social Class and Capital

There have been a number of changes in capitalism in this current period of neoliberal globalization. One development is the growth in service, communications, and technological industries in the developed world. One such "service industry" is education, for as the International Chamber of Commerce (ICC) observes, "services are coming to dominate the economic activities of countries at virtually every stage of development" (ICC, 1999, p. 1).

Another development is the declining profitability of capital—the crisis of capital accumulation. This crisis has resulted in intensification of competition between capitals, between national and transnational capitals and corporations. There is general agreement among critical educators and Marxists that "the pressure on nations to liberalize services at the national level can be seen, therefore, as a response to the declining profitability of manufacture" (Beckmann and Cooper, 2004). This crisis of capital accumulation, as predicted by Marx and Engels (1848), has led to the intensification of the extraction of surplus value, the progressing global immiseration of workers, and the intensifying control of populations by the ideological and repressive state apparatuses identified and analyzed by Althusser.[1]

Class War from Above

Neoliberal and neoconservative policies aimed at intensifying the rate of capital accumulation and extraction of surplus value comprise an intensification of "class war from above" by the capitalist class against the working class. One major aspect of this is the fiscal policy of increasing taxes on workers and decreasing taxes on business and the rich. Of course, some people don't like trillion dollar tax handouts to the rich. These oppositionists have to be denigrated, scorned, and controlled! This is where neoconservative policies are important. On the one hand they persuade the poor to vote (right-wing Republican) for a social or religious or antiabortion or homophobic or racist agenda against their own (more left-wing, more Democrat, or further Left) economic self-interest.

The class war from above, therefore, has a neoliberal, economic element, but it has also embraced a neoconservative political element to strengthen the force of the state behind it. In Andrew Gamble's words, it is *The Free Economy and the Strong State* (1988), a state strong on controlling education, strong on controlling teachers, strong on marginalizing oppositional democratic forces such as local elected democracy, trade unions, critical educators, critical students. Moreover, neoconservatism aids in the formation of a state strong on enforcing the neoliberalization of schools and society.

Despite the horizontal and vertical cleavages within the capitalist class (Dumenil and Levy, 2004), the architects of neoliberal and neoconservative policies know very well who they are. Nobody is denying capitalist class consciousness. They are rich. They are powerful. And they are transnational as well as national. They exercise (contested) control over the lives of worker-laborers and worker-subjects. If there is one class that does not lack class-consciousness, *the subjective appreciation of its common interest, and its relationship within the means of production to other social classes*, it is the capitalist class.

Members of the capitalist class do recognize that they survive in dominance *as a class* whatever their skin color, or dreams, or multifaceted subjectivities and histories of hurt and triumph; they survive precisely because they do know they are a class. They have class consciousness, they are "a class *for* themselves" (a class with a consciousness that they are a class), as well as a "class *in* themselves" (a class or group of people with shared economic conditions of existence and interests). The capitalist class does not tear itself to pieces negating or suborning its class identity, its class awareness, its class power over issues of "race" and gender (or, indeed, sexuality or disability). And they govern in their own interests, not just in education "reform" but also in enriching and empowering themselves while disempowering and impoverishing others, namely, the (white and black and other minority, male and female) working class.

The Growth of National and Global Inequalities

Inequalities both between states and within states have increased dramatically during the era of global neoliberalism. Global Capital, in its current neoliberal form in particular, leads to human degradation and inhumanity and increased social class inequalities within states and globally. These effects are increasing (racialized and gendered) social class inequality within states, increasing (racialized and gendered) social class inequality between states. The degradation and capitalization of humanity, including the environmental degradation, impact primarily in a social class–related manner. Those who can afford to buy clean water do not die of thirst or diarrhea. In many states across the globe, those who can afford school or university fees, where charges are made, end up without formal education or in grossly inferior provision. Hearse (2009) points out that the Golden Age for the salaried worker across all the OECD countries was between 1945 and 1973, when ordinary working people gained their highest percentage share of GDP. Since then the real wages of the middle and working class have stagnated or fallen, while income for the rich has rocketed and that of the super-rich has hit the stratosphere, Hearse continues, the facts are astounding. Contrary to the delusions of the free-market fundamentalists, the Thatcher/Reagan revolution has come at a great cost to the working and middle classes. In the United States, the top 1 percent have seen a 78 percent increase in their share of national income since 1979 with the bottom 80 percent of the population experiencing a 15 percent fall. The current form of globalization is tightening rather than loosening the international poverty trap. Living standards in the least developed countries are now lower than they were thirty years ago. Inequalities within states have widened partly because of the generalized attack on workers' rights and trade unions, with restrictive laws passed hamstringing trade union actions (Rosskam, 2006. See also Hill et al, 2006; Hill, 2009a, b; Hill and Kumar, 2009; Hill and Rosskam, 2009). And it is workers now being asked to pay for the crisis. Under capitalism, it usually is. It is workers and their trade unions voluntarily, or under pressure, accepting cuts in pay and conditions. It is workers and their families in the advanced capitalist world whose children will pay back the state for the "the bankers' bailout," the trillions of dollars handed to industrial and finance capital.

Increasingly Unequal Distribution of Wealth in the United States and Britain

David Harvey (2005) argues that while the intellectual origins of neoliberalism reach back to the 1930s, its material origins stem from the crisis of capital accumulation of the late 1960s and 1970s. In his estimation, this crisis constituted both a political and an economic threat to economic elites and ruling classes across both the advanced capitalist countries and the developing countries (p. 15). In the United States, prior to the 1970s, the wealthiest 1 percent of the population owned between 30 and 47 percent of all wealth assets (p. 16). But in the 1970s it slid to just 20 percent. Asset values collapsed. In Harvey's phrase, "the upper classes had to move decisively if they were to protect themselves from political and economic annihilation" (ibid.). And they did, leading Harvey to conclude that we can best understand neoliberalization as a project designed to achieve the restoration of class power. Furthermore, given that by 1998 the percentage ownership of all wealth assets in the United States held by the wealthiest 1 percent of the population had almost doubled since the mid-1970s, we should view the neoliberal project as having achieved great success.

Likewise, in the United Kingdom, the wealth of the superrich doubled since Tony Blair came to power in 1997. According to the Office for National Statistics (2000), nearly six hundred thousand individuals in the top 1 percent of the U.K. wealth league owned assets worth £355 billion in 1996, the last full year of Conservative rule. By 2002, that had increased to £797 billion.

Increasingly Unequal Distribution of Income in the United States and Britain

As for income, the ratio of the salaries of chief executive officers (CEOs) to the median compensation of workers increased from just over thirty to one in 1970 to nearly five hundred to one by 2000 (Harvey, 2005, p. 17). Korten (2004) highlights the immense increase in salaries taken by top U.S. executives since the early 1990s. In the United States, between 1990 and 1999 inflation increased by 27.5 percent, workers' pay by 32.3 percent, corporate profits by 116 percent, and, finally, the pay of chief executive officers by a staggering, kleptocratic, 535 percent (p. 17; see also Brenner, 2005).

In the United States, the share of the national income taken by the top 1 percent of income earners had been 16 percent in the 1930s. It fell to around 8 percent between the mid-1940s and mid-1970s. The neoliberal revolution restored its share of national income to 15 percent by the end of the twentieth century. And the Federal minimum wage, which stood on a par with the poverty level in 1980, had fallen to 30 percent below that in 1990: "The long decline in wage levels had begun" (Harvey, 2005, p. 25). Pollin (2003) shows that, in the United States, the level of "real wages" per hour dropped from $15.72 in 1973 to $14.15 in 2000. In the United Kingdom, the top 1 percent of income earners have doubled their share of the national income from 6.5 to 13 percent since 1982 (Harvey, 2005, pp. 15–18). Tax policy has been crucial in affecting these growing inequalities.

Changing Tax Rates: Capitalist Winners and Working Class Losers

Dumenil and Levy (2004) show that in the United States, those in the highest tax bracket are paying tax at a rate around half that of the 1920s, whereas the current tax rate for those in the lowest tax bracket has more than doubled over the same period. As a forerunner of George W. Bush's "trillion dollar tax giveaway to the rich," Ronald Reagan cut the top rate of personal tax from 70 to 28 percent. Both the Reagan and Thatcher governments also dramatically cut taxes on business/corporations.

In Britain, too, the working class is paying more tax. The richest groups are paying a smaller proportion of their income in taxes in comparison to 1949 and to the late 1970s. These dates were both in the closing stages at the end of two periods of what might be termed "Old Labour," or social democratic governments (in ideological contradistinction to the primarily neoliberal policies of "New Labour"). As Paul Johnson and Frances Lynch reported in their 2004 article in *The Guardian*, in comparison with the late 1970s, the "fat cats" are now paying around half as much tax (income tax and insurance contribution rate). These fat cats are paying less income tax and national insurance as a percentage of their earned income than in 1949. "As a percentage of income, middle and high earners pay less tax now than at any time in the past thirty years." (Johnson and Lynch, 2004). In contrast, the average tax-take for "the low paid" (allowing for inflation) is roughly double that of the early 1970s—and nearly twice as much as in 1949 (Johnson and Lynch, 2004). No wonder, then, that Johnson and Lynch titled their article: "Sponging off the Poor."

Capital, Corporations, and Education

Education is now big business—"edu-business." Current worldwide spending in education is "estimated at around 2,000 billion dollars...more than global automotive sales" (Santos, 2004, p. 17). According to Santos, "capital growth in education has been exponential, showing one of the highest earning rates of the market: £1000 invested in 1996 generated £3,405 four years later" (pp. 17–18, cited in Delgado-Ramos and Saxe- Fernandez, 2005). Santos continues, "that is an increased value of 240%, while the London Stock Change valorization rate accounted on the same period for 65%. Other 2004 data indicate that, current commercialized education, incomplete as it is, already generates around $365 billion in profits worldwide" (ibid.).

Capital—national and transnational corporations along with their major shareholders—has a number of plans with respect to education. First, there is "The Capitalist Plan *For E*ducation." This plan aims to produce and reproduce a workforce and citizenry and set of consumers fit for capital. According to this plan, schools must serve two overriding functions, an ideological function and a labor training function. These comprise socially producing labor power for capitalist enterprises. This is people's capacity to labor—their skills and attitudes, together with their ideological compliance and suitability for capital—as workers, citizens, and consumers. In this analysis, Althusser's concepts (1971) of schools as ideological state apparatuses (ISAs) is useful, with schools acting as key elements in the ideological indoctrination of new citizens and workers into thinking "there is no alternative" to capitalism, that capitalism and competitive individualism with gross inequalities is "only natural" (see also Hill, 2001, 2003, 2004b, 2006b, 2007).

Second, there is "The Capitalist Plan *In* Education," which entails smoothing the way for direct profit-taking/profiteering from education. This plan is about how capital wants to make *Direct Profits* from education. This centers on setting business "free" into education for profit-making and profit-taking—extracting profits from privately controlled/owned schools and colleges or from aspects of their functioning. Common mechanisms for such profiteering include: managing, advising, controlling, and owning schools. These possibilities are widened in the United Kingdom by New Labour's Education White Paper of October 2005.

Finally, there is "Capital's Global Plan for Education Corporations." This is a series of national capitalist plans for domestically based national or multinational corporations globally. This is a plan for U.K., U.S., Australia, New Zealand, and locally (e.g., in particular states such as Brazil in Latin America) based "edu-businesses" and corporations to profit from international privatizing, franchising, and marketing activities. With a worldwide education industry valued at two trillion dollars annually, "it is not surprising that many investors and 'edupreneurs' are anxious to seize the opportunities to access this untapped gold mine" (Schugarensky and Davidson-Harden, 2003, p. 323). It is not just national edu-businesses that are involved—it is large multi-activity national and global capitalist companies as well.

The restructuring of education has taken place/is taking place throughout the globe. Neoliberalization, accompanied by neoconservative policies (Dumenil and Levy, 2004; Harvey, 2005), has proceeded apace, spurred by governments committed to developing human capital and labor power better suited to the interests of capital and the owners of capital, the capitalist class. This restructuring is also developing and promoting the capitalist class's own edu-businesses in order to gain these service exports, the export of educational services (Hatcher, 2001; Rikowski, 2002).

Internationally, liberalization of schooling and higher education, as well as other education sectors, has either been taken up voluntarily or been forced upon governments through the influence of the world regime of neoliberal capitalist organizations such as the World Bank, the International Monetary Fund (IMF), OECD, international trade regimes such as the WTO's General Agreement on Trade in Services

(GATS) and the proposed Free Trade Area of the Americas (FTAA), and regional derivatives/government/national capital/U.S. capital hemispheric organizations such as the Partnership for Educational Revitalization in the Americas (PREAL) in Latin America. Where world or regional organizations of capital are not successful in implementing liberalization, local free trade agreements (FTAs) and bilateral intergovernmental agreements are opening up "free trade" in services such as "education services."[2]

Neoliberal Policies

There are a number of common aspects of the neoliberalization of schooling and education services (see Giroux, 2004; Harvey, 2005; Ross and Gibson, 2007; Hill, 2009a). It is possible to identify twelve aspects of neoliberal policy within states, and a further four in terms of global policy. Within states, these are as follows. The first policy is low public expenditure. Typically there has been a regime of cuts in the postwar Welfare State, the withdrawal of state subsidies and support, and the transition toward lower public expenditure. This has involved public expenditure cuts in education, driven primarily and most significantly by an economic imperative to reduce aggregate social expenditures. In developed states this has been termed "prudence" or "sound fiscal policy." In developing and less developed states, this policy has been a condition of structural adjustment programs and loans (SAPs and SALs) administered by the World Bank and IMF (Chossudovsky, 1998) designed primarily with debt servicing obligations in mind.

Within national economies there have been both the policies of controlling inflation by interest rates, preferably by an independent central bank, policies responsive to the needs of capital rather than responsive to domestic political demands, and the policy of balancing budgets instead of using budget imbalances to stimulate demand.

A salient policy development is privatization of formerly publicly owned and managed services. Variously termed "liberalization" (e.g., by the International Labour Organization) and neoliberalization, this comprises transferring, selling off, the means of production, distribution, and exchange, and also, in the last two decades globally, services such as education and health into private ownership.

Policies that have served to "soften up" public opinion and service provision for privatization include the setting up of markets (or quasi-markets) in services and competition between different "providers," such as universities and schools, for (high potential) students/pupils.

A concomitant of marketization is decentralization: in general, neoliberal education policies, for example, in Latin America (Carnoy, 2002) and elsewhere (such as in England and Wales), have taken the shape of "decentralization" efforts, aimed at scaling down the role of central governments in direct responsibility for different aspects of education, toward increased provincial/regional, municipal and private involvement in education.

These policies of privatization, fiscal "rectitude," decentralization, and deregulation commonly result in increasingly differentiated provision of services. Within states this results in intensified hierarchical differentiation between educational institutions on the basis of ("raced" and gendered) social class.

Within states, neoliberal education policies stress selective education. Within education, whether through the development of private schools and universities or through the creation of different (and hierarchically arranged) types of schools and universities (as in Britain), the public right to education has been transformed into the creation of "opportunity" to acquire the means of education and additional cultural capital through selection, through a selective and hierarchically stratified schooling and education system.

There is increasing differentiation globally too. Neoliberalization of schooling services, in particular higher education, has reinforced the relegation of most developing states and their populations to subordinate global labor market positions, specializing in lower skilled services and production. This global differentiation is enforced by the World Bank and other international agency prescriptions regarding what education should be provided, and at what levels, in less developed and developing states (Leher, 2004).

Schools and universities are increasingly run in accordance with the principles of "new public managerialism" (Mahoney and Hextall, 2003) based on a corporate managerialist model imported from the world of business. As well as the needs of capital dictating the principal aims of education, the world of business also supplies the model of how it is to be provided and managed.

A key element of capital's plans for education is to cut its labor costs. For this, a deregulated labor market is essential—with schools and universities able to set their own pay scales and sets of conditions—busting national trade union agreements and weakening union powers to protect their workforces. Thus, where liberalism reigns there is relatively untrammeled selling and buying of labor power, resulting in a "flexible," poorly regulated labor market (Costello and Levidow,

2001). Some impacts on workers' rights, pay, and what the International Labour Organization calls "securities" are spelt out here.

Internationally, neoliberalism requires untrammeled free trade. Currently the major mechanism for this is the GATS, though there are and have been many other mechanisms.[3] One aspect of this free trade is that barriers to international trade, capitalist enterprise, and the extraction of profits should be removed. This applies as much to trade in services such as education and health as it does to the extraction of oil or the control of water supply.

There should also be a "level playing field" for companies of any nationality within all sectors of national economies. Barriers such as "most favored nation" (MFN) clauses should be dismantled, allowing any corporation, whether domestic or foreign or transnational, to own/run universities, teacher education, schooling in any state. Neoliberal capital also demands that international trade rules and regulations are necessary to underpin free trade, with a system for penalizing "unfair" trade policies such as subsidies and the practice of promoting/favoring national interests or national workforces such as teachers and lecturers. Certainly, according to the GATS, this level playing field will be legally enforceable under pain of financial penalties for any state that has signed up for education services particular to the GATS.

There is an exception to these free trade demands by transnational capital. The earlier-mentioned restrictions do not apply in all cases to the United States (or other major centers of capitalist power such as the European Union). Ultimately, the United States may feel free to impose "economic democracy" and "choice" by diplomatic, economic, or military means. Ultimately peoples and states can be coerced to choose, bombed to obey.

Thus, key strategies to maximize capital accumulation, and to increase the rates of profit, are global free trade and privatization within states. Hirtt (2004) summarizes the "New Economic Context" (of neoliberalism) as having four characteristics:

- The intensification ("globalization") of economic competition.
- A decrease in state financial resources for public sector provision such as school or university education.
- A faster pace of change (with rapid developments in technology and in opening up new markets).
- And a "polarization" of the labor market—with less being spent on the education of "the masses" in particular.

Two of Hirtt's characteristics are contextual (increasing competition and faster technology/opening up new markets). In the other two (public expenditure cuts, and increasing polarization/hierarchicalization of education and the workforce) Hirtt identifies what he considers the salient intentions and effects of neoliberal policies.

He also draws attention to a seeming contradiction between two of these four intentions, where:

> the industrial and financial powers ask the political leaders to transform education so that it can better support the competitiveness of regional, national or European companies. But, on the other hand, the same economic powers require that the State reduce its fiscal pressure and thus reduce its expenditure, notably in the field of education. (pp. 444–445)

The contradiction is solved by polarization. The poor (in general) are polarized to the bottom of an intellectually and materially worsening education. Why educate them expensively?

Neoconservative Policies

There are, of course, resistant teachers, teacher educators, students, and student teachers who seek better and more hopeful pasts, presents, and futures rooted in experiences and histories predating and seeking to postdate the combined neoliberal and neoconservative storming of the ramparts of the state and the education state apparatuses—ministries, schools, vocational colleges, and universities.

Much of this chapter is about how capital, and the governments and state apparatuses serving their interests, "get away with it," fight the "culture wars," and seek to attain ideological hegemony for neoliberalism while displacing oppositional counter-hegemonic liberal-progressive, Marxist/socialist, and social democratic ideals and "common sense." This partly takes place through a process of a systematic denigration and humbling of publicly provided services and public sector workers as bureaucratized, slow to adapt, resistant to change, expensive, and putting their own interests above that of the service and of the "consumers" of those services. It also takes place through conservative control of the curriculum and pedagogy that seeks to silence or discredit or marginalize counter-hegemonic ideologies.

A policy that is both neoliberal and neoconservative—partly aimed at whipping these resistant and critical students, teachers, professors in line—is employment policy. Enforcing acceptance of the neoliberal revolution and weakening opposition to it is partly carried out through the importation of "new public managerialism" into the management of schools and colleges and education services. This involves the surveillance of teachers and students, partly through the imposition of tightly monitored testing of chunks of knowledge deemed by national and state/local governments to be suitable, and sanitized, and conservative enough. Conservatism is enforced through the curriculum and the SATS (Standard Assessment Tests). Indeed, in England and Wales, Conservative prime minister Margaret Thatcher personally intervened on a number of occasions during the drawing up of the various subject outlines for the national curriculum of 1988—a curriculum largely maintained by New Labour today (Hill, 2006).

There are three major aspects of neoconservatism. The first is that of the circumscription described earlier, the attempt to straightjacket students', teachers', and professors' practices—their curricula, their pedagogy, and their use of their time in class and for homework. This is the repressive use of the local state apparatus.

The second is the degree of enforcement by the central state apparatuses, including those of the security state. This includes blacklists, non-promotion of oppositional teachers and professors, and public vilification such as the right-wing campaign against "The Dirty Thirty" left-wing professors at UCLA (Weiner, 2006). There is not yet a resuscitation of the McCarthyite House of UnAmerican Activities Committee, but the widespread antiterror legislation such as the so-called PATRIOT Act in the United States, including the right of the security services to track the library borrowing habits of U.S. residents, does serve to diminish oppositional activity.

The third aspect of neoconservatism is the ongoing "culture wars," the use of the ideological state apparatuses (some churches, many schools, nearly all mass media) to legitimate neoliberal and neoconservative ideology and "common-sense" practices and beliefs. Although there is the appearance of ideological choice between, for example, the major political parties in the United States and Britain, or between mass circulation television and newspapers, seriously oppositional views are erased from television programs and party platforms. People who question the "tweedledum and tweedledee" choice in politics and the media tend to be regarded in much of the United States and Britain (though not all) as suspect—and can be transferred from the ministrations of the ideological state apparatuses to the attentions of the repressive state apparatuses.

A recent example in England and Wales is the debate about the New Labour government's October 2006 White Paper on Education (Her Majesty's Government, 2005). This is a major step toward the neoliberalization of state schooling in England and Wales in terms of marketization and pre-privatization (which seeks to further open up schools to private and business ownership, control, and sponsorship, including setting up a system of new "Independent Trust Schools," state primary and secondary schools that can become self-governing in terms of finances, admissions, curriculum and staffing; Rikowski, 2005a, b).[4] This has occasioned vigorous debate and fairly unprecedented opposition from within the Labour Party itself (e.g., through the pressure group, Compass).[5]

However, the debate is largely defensive, seeking largely to defend the continued role of local education authorities and opposing further marketization and neoliberalization of schools. It is scarcely proactive. It does not seek the reversal of the New Labour policies since 1997 on marketization, nor the changing of the national curriculum (Hill, 2006). Similarly with teacher education, in virtually all current discussion about the curriculum for "teacher training" in England and Wales, there is an acceptance of the status quo, substantially introduced as part of the Thatcher-Major revolution in education. Deep critique of the teacher training curriculum is rare.

Neoliberalism and Neoconservatism: Global Similarities, National Variations

While there are global similarities in liberalizing education policy, there are national and also local variations in the type and extent of the various policies (see Hill et al., 2006; Jones, Cunchillos, Hatcher and Hirtt, 2007; Hill, 2009a, b; Hill and Kumar, 2009; Hill and Rosskam, 2009). These relate to different historical conditions and balance of forces—the relative strengths of the trade union movement, workers' trade union, and political organizations, and other forces in Civil Society, with their varying strengths of resistance to neoliberal policies, on the one hand, and of local Capital on the other. We are not in an era of the unimpeded march to neoliberal capitalism. Comparing three North American states, for example (Canada, the United States, and Mexico), shows some similarities and some differences in context and policy.[6]

Nor are we in an era of the unimpeded march of neoconservatism. Western Europe in general fails to comprehend the specific religious right and radical right agenda in the United States. Such appeals go relatively unheeded, and meet with incomprehension not just in social democratic Scandinavia but in the rest of Western Europe, too. In Western Europe, outside of wartime, economic issues tend to prevail in elections. A partial exception is currently in Britain, where all three parties, New Labour, Liberal Democrat, and Conservative, are supporting similar versions of "compassionate" neoliberalism, and where New Labour and Conservative appear identical in their neoconservatism. However, historically in Britain, and currently in the rest of Western Europe, parties are more social-class-based and have economic agendas displaying far more divergence than in the United States and Britain today.

The Impacts So Far

Capital has not merely developed these plans and set them on the shelf for future reference. Many elements of these plans have already been put into action. We can already see much of the impact of these initiatives in the growing inequities they produce in the lives of students, declining levels of democratic control over schools, and worsening work conditions in the teaching and other education professions.

Impacts on Equity: Neoliberalism and Widening ("Raced" and Gendered) Class Inequalities in Education

Where there is a market in schools (where high status schools can select their intakes, either based on "academic achievement" or on other class-related criteria such as "aptitudes"), the result is increasing raced and gendered social class differentiation. The middle classes (predominantly white) rapidly colonize the "best" schools; the working classes (white and black) get pushed out. They don't get through the school gate. High status/high achieving middle class schools get better and better results. In a competitive market in schools, "Sink" schools sink further, denuded of their "brightest" intakes.

The same is true of higher education. In the United States it is highly tiered—there is a hierarchy. Entry to elite schools/universities is very largely dependent on a student's ability to pay—on social class background. This is intensifying in England and Wales too. Until the 1980s, there were no university fees in Britain—the state/the taxpayer paid. Entry was free for students. For the last twenty year, all universities charged the same for undergrad courses. Now, the New Labour government is introducing "variable" fees for different universities. Britain is "going American." This will reinforce elitism and exclude poorer groups, especially minorities, but white working class students, too.

Neoliberalization of schooling and university education is accompanied by an increase in (gendered, raced, linguistically differentiated) social class inequalities in educational provision, attainment, and subsequent position in the labor market. For example, the movement to voucher and charter schools as well as other forms of privatized education such as chains of schools[7] have proven to be disproportionately beneficial to those segments of society who can afford to pay for better educational opportunities and experiences, leading to further social exclusion and polarization.[8]

Hirtt (2004) has noted the apparently contradictory education policies of capital, "to adapt education to the needs of business and at the same time reduce state expenditure on education" (p. 446). He suggests that this contradiction is resolved by the polarization of the labor market, that from an economic point of view it is not necessary to provide high level education and general knowledge to all future workers. "It is now possible and even highly recommendable to have a more polarized education system...education should not try to transmit a broad common culture to the majority of future workers, but instead it should teach them some basic, general skills" (ibid.).

In brief, then, manual and service workers receive a cheaper and inferior education limited to transferable skills while elite workers receive more expensive and more and internationally superior education. Not only does this signal one manifestation of the hierarchicalization of schools and the end of the comprehensive ideal, but it also represents a form of educational triage—with basic skills training for millions of workers, more advanced education for supervision for middle class and in some countries the brightest of the working classes, and elite education for scions of the capitalist and other sections of the ruling classes.

Impacts on Workers' Pay, Conditions, and Securities

A key element of Capital's plans for education is to cut its labor costs. For this, a deregulated labor market becomes essential—with schools and colleges able to set their own pay scales and sets of conditions—busting national trade union agreements, and, weakening the power of trade unions—such as teacher unions—to protect their workforces. As a consequence of capital's efforts to extract higher rates of surplus value from their labor power, educational workers suffer declining pay, decreasing benefits, and deteriorating working conditions (Hill, 2005b). There is the ongoing *casualization* of academic labor, and the increased *proletarianization* of the teaching profession.

By "casualization," I mean the move toward part-time and temporary employment in the education sector. Simultaneously, the "proletarianization" of teaching results in:

- declining wages, benefits, and professional autonomy for teachers;
- growing intensification of teachers' labor through increases in class-sizes and levels of surveillance; and
- mounting efforts to eliminate the influence of teachers' unions as mechanisms for promoting and defending teachers' interests.

The intensification of work is justified in different countries through campaigns of vilification against public service workers such as teachers and education officials. Siqueira (2005) reports that in Brazil, the Cardoso government of the mid-1990s launched, using the media, a renewed and stronger campaign against civil servants, unions, and retired public employees.

Some of the usual terms used by his government to refer to these groups were: sluggish, negligent, agitators, old-fashioned, unpatriotic, selfish, and lazy. This is part of the global neoliberal critique of public service workers for being expensive self-interested workers who have "captured" the professions with their restrictive and expensive practices. In Britain, Stephen Ball (1990) has called this denigration, "a discourse of derision" (p. 22). In some right-wing newspapers, such as *The Daily Mail* in Britain, it is more like a "discourse of hate." One need only recall former secretary of education Rod Paige's denigration of the National Education Association as a "terrorist organization" to find a potent example of such speech in the United States.

Impacts on Democracy and on Critical Thinking

The neoconservative faces of education reform, indeed, of the wider marketization and commodification of humanity and society, come to play in the enforcement and policing of consent, the de-legitimizing of deep dissent, and the weakening of oppositional centers and practices and thought. In eras of declining capital accumulation, an ultimately inevitable process, capital—and the governments and parties and generals and CEOs who act at their behest—more and more nakedly ratchet up the ideological and repressive state apparatuses of control (see also Hill, 2001, 2003, 2004b, 2006b, and 2007). Thus, key working class organizations such as trade unions and democratically elected municipal governments are marginalized, and their organizations, and those of other radically oppositional organizations based on race, ethnicity, religion, are attacked through laws, rhetoric, and, ultimately, sometimes by incarceration. In education, the combined neoliberal-neoconservative educational reform has led to a radical change in what governments and most school and college managements/leaderships themselves see as their mission. In the 1960s and 1970s (and with long prior histories), liberal-humanist or social democratic or socialist ends of education were common through the advanced capitalist (and parts of the anticolonialist developing) worlds.

This has changed dramatically within the lifetimes of those over thirty. Now the curriculum is conservative and it is controlled. Now the hidden curriculum of pedagogy is performative processing and "delivery" or pre-digested points. Now the overwhelming and nakedly overriding and exclusive focus is on the production of a differentially educated, tiered (raced and gendered) social class workforce and compliant citizenry. Differentially skilled and socially/politically/culturally neutered and compliant human capital is now the production focus of neoliberalized education systems and institutions, hand in glove with and enforced by a Neoconservative ideology and state.

Resistance

But there is resistance; there are spaces, disarticulations, and contradictions (see for example, Jones, Cunchillos, Hatcher and Hirtt, 2007; and Hill, 2009b). There are people who want to realize a different vision of education. There are people who want a more human and more equal society, a society where students and citizens and workers are not sacrificed on the altar of profit before all else.

And there are always, sometimes minor, sometimes major, awakenings that the material conditions of existence, for teacher educators, teacher, students, and workers and families more widely, simply do not match or recognize the validity of neoliberal or neoconservative or other capitalist discourse and policy.

Cultural Workers as Critical Egalitarian Transformative Intellectuals and the Politics of Cultural/Educational Transformation

What influence can critical librarians, information workers, cultural workers, teachers, pedagogues have in working toward a democratic, egalitarian society/economy/polity?[9] How much autonomy from state suppression and control do/can state apparatuses and their workers—such as librarians, teachers, lecturers, youth workers, have in capitalist states such as England and Wales, or the United States? Don't they get slapped down, brought into line, controlled, or sat upon when they start getting dangerous, when they start getting a constituency/having an impact? When their activities are deemed by the capitalist class and the client states and governments of/for capital to be injurious to the interests of (national or international) capital?

The repressive cards within the ideological state apparatuses are stacked against the possibilities of transformative change through the state apparatuses and their agents. But historically and internationally, this often has been the case. Spaces do exist for counter-hegemonic struggle—sometimes (as in the 1980s and 1990s) narrower, sometimes (as in the 1960s and 1970s and currently) broader. By itself, divorced from other arenas of progressive struggle, its success, the success of radical librarians, cultural workers, media workers, education workers will be limited.

This necessitates the development of proactive debate both by and within the Radical Left. But it necessitates more than that; it calls for direct engagement with liberal, social democratic, and Radical Right ideologies and programs, including New Labour's, in all the areas of the state and of civil society, in and through all the ideological and repressive state apparatuses, and in and through organizations and movements seeking a democratic egalitarian economy, polity, and society.

It takes courage, what Gramsci called "civic courage." It is often difficult. Some of our colleagues/comrades/companeras/companeras/political and organizational coworkers ain't exactly easy to get along with. Neither are most managements; especially those infected with the curse of "new public managerialism," the authoritarian managerialist, brutalist style of management and (anti-) human relations, where "bosses know best" and "don't you dare step outa line, buddy!"

But I want here to modify the phrase "better to die on your feet than live on your knees." It is of course better to live on your/our feet than live on your/our knees. And whether it is millions on the streets defending democratic and workers' rights (such as over pensions, in Britain and elsewhere, or opposing state sell-offs of publicly owned services, in France and elsewhere, or laws attacking workers' rights, in Italy and Australia and elsewhere)—all in the last two years—or in defense of popular socialist policies in Venezuela, Bolivia, Honduras, Nepal, we are able, in solidarity, and with political aims and organization, not only to stand/live on our feet, but to march with them, to have not just an individual impact, but a mass/massive impact. We have a three-way choice—to explicitly support the neoliberalization and commodification and capitalization of society; to be complicit, through our silence and inaction, in its rapacious and anti-human/antisocial development, or to explicitly oppose it. To live on our feet and use them and our brains, words, and actions to work and move with others for a more human, egalitarian, socially just, economically just, democratic, socialist society: in that way we maintain our dignity and hope.

Notes

1. See Althusser, 1971. See also Hill, 2001, 2004b, 2005b.
2. See Hill, 2005b; and Hill et al., 2006; Leher, 2004; Rikowski, 2002, 2003; and Schugarensky and Davidson-Harden, 2003.
3. See Devidal, 2004; Schugarensky and Davidson-Harden, 2003; and Sinclair and Grieshaber-Otto, 2002.
4. See Hatcher (2005a and b) for a contrasting Marxist analysis that suggests that control over the reproduction of labor power is more salient than a pre-privatization agenda. Hatcher's argued view is that the view of British capital is that the most favorable conditions for the production of "human capital" for the economic competitiveness of British capital are best secured by the state directly providing school education. This contrasts with the views expressed on this matter by Rikowski and by Hill.
5. See the Compass website at http://www.compassonline.org.uk/about.asp.
6. Hursh (2005) examines differences and similarities between the neoliberal education agenda in England and Wales and the United States. And in Hill (2005b) and Hill et al. (2006) I attempt to show global similarities.
7. See Molnar, 2001 and 2005; and Molnar et al., 2004
8. See Gillborn and Youdell, 2000; Hill, Maisuria, and Greaves, 2006; Hill et al., 2006; Whitty et al., 1998.
9. For a fuller discussion of the concept of teachers as transformative intellectuals, or as revolutionary critical intellectuals, see, e.g., Giroux, 1988; Hill, 1989, 2003, 2005c; McLaren, 2000, 2005, and 2006; McLaren and Farahmandpur, 2005; and McLaren and Rikowski, 2001.

References

Althusser, L. (1971). Ideology and state apparatuses. In L. Althusser (ed.). *Lenin and Philosophy and Other Essays.* London: New Left Books.

Ball, S. (1990). *Politics & Policymaking in Education.* London: Routledge.

Beckmann, A., and Cooper, C. (2004). "Globalization," the new managerialism and education: rethinking the purpose of education. *Journal for Critical Education Policy Studies*, 2(1). Available at http://www.jceps.com/index.php?pageID=article&articleID=31.

Brenner, R. (2005). *The Economics of Global Turbulence.* London: Verso.

Carnoy, M. (2002). Latin America: the new dependency and educational reform. In H. Daun (ed.). *Educational Restructuring in the Context of Globalization and National Policy.* New York: RoutledgeFalmer.

Chossudovsky, M. (1998). *The Globalization of Poverty: Impacts of IMF and World Bank Reform.* Halifax: Fernwood.

Costello, A., and Levidow, L. (2001). Flexploitation strategies: UK lessons for Europe. *The Commoner,* 1. Available at http://www.commoner.org.uk/Flex3.pdf.

Delgado-Ramos, G., and Saxe-Fernandez, J. (2005). The World Bank and the privatization of public education: a Mexican perspective. *Journal for Critical Education Policy Studies,* 3(1). Available at http://www.jceps.com/index.php?pageID=article&articleID=44#sdfootnote10sym.

Devidal, P. (2004). Trading away human rights? The GATS and the right to education. *Journal for Critical Education Policy Studies,* 2(2). Available at http://www.jceps.com/index.php?pageID=article&articleID=28.

Dumenil, G., and Levy, D. (2004). *Capital Resurgent: Roots of the Neoliberal Revolution.* London: Harvard University Press.

Gamble, A. (1988). *The Free Economy and the Strong State: The Politics of Thatcherism.* London: MacMillan.

Gillborn, D., and Youdell, D. (2000). *Rationing Education: Policy, Practice, Reform and Equity.* Buckingham, UK: Open University Press.

Giroux, H. (1988). *Teachers as Intellectuals: Toward a Critical Pedagogy of Learning.* Granby, Massachusetts: Bergin and Garvey.

———. (2004). The *Terror of Neoliberalism: Authoritarianism and the Eclipse of Democracy.* Boulder, CO: Paradigm.

Harvey, D. (2005). *A Brief History of Neoliberalism.* Oxford, England: Oxford University Press.

Hatcher, R. (2001). Getting down to the business: schooling in the globalised economy. *Education and Social Justice,* 3(2), 45–59.

———. (2005a). *The White Paper: What Does It Intend, What Would It Mean, Will It Happen? Some Initial Thoughts on Some of the Themes in Labour's Latest Policy Document*, unpublished paper, November 4. Available in Word by email attachment from Rikowskigr@aol.com.

———. (2005b). *Business Sponsorship of Schools: For-Profit Takeover or Agents of Neoliberal Change? A Reply to Glenn Rikowski's "Habituation of the Nation: School Sponsors as Precursors to the Big Bang?"* November 5, available on Glenn Rikowski's web log: *The Volumizer*, at: http://journals.aol.co.uk/rikowskigr/Volumizer/entries/651, posted November 7. Also available from the MASSES (Marxian Analysis of Schools, Society, and Education SIG) list, posted November 5, at: http://groups.yahoo.com/

group/MarxSIG. Available also in Word by email attachment, from: Rikowskigr@aol.com.

Hearse, P. (2009). Has working class consciousness collapsed? *International Viewpoint.* Online at http://www.internationalviewpoint.org/spip.php?article1516.

Her Majesty's Government. (2005). *Higher Standards, Better Schools for ALL—More Choice for Parents and Pupils.* White Paper, Cm6677, October. Norwich: The Stationery Office.

Hill, D. (1989). *Charge of the Right Brigade: The Radical Right's Assault on Teacher Education.* Brighton: Institute for Education Policy Studies. Available at http://www.ieps.org.uk.cwc.net/hill1989.pdf.

———. (2001). State theory and the neoliberal reconstruction of schooling and teacher education: a structuralist neo-Marxist critique of postmodernist, quasi-postmodernist, and culturalist neo-Marxist theory. *British Journal of Sociology of Education*, 22(1), 137–157.

———. (2003). Global neo-liberalism, the deformation of education and resistance. *Journal for Critical Education Policy Studies,* 1(1). Available at http://www.jceps.com/index.php?pageID=article&articleID=7.

———. (2004b). Books, banks and bullets: controlling our minds—the global project of imperialistic and militaristic neo-liberalism and its effect on education policy. *Policy Futures,* 2(3). Available at http://www.triangle.co.uk/pfie/.

———. (2005a). State theory and the neoliberal reconstruction of schooling and teacher education. In G. Fischman, P. McLaren, H. Sünker, and C. Lankshear (eds.). *Critical Theories, Radical Pedagogies and Global Conflicts.* Boulder: Rowman &Littlefield.

———. (2005b). Globalisation and its educational discontents: neoliberalisation and its impacts on education workers' rights, pay, and conditions. *International Studies in the Sociology of Education*, 15(3), 257–288.

———. (2005c). Critical education for economic and social justice. In M. Pruyn and L. Huerta-Charles (eds.). *Teaching Peter McLaren: Paths of Dissent.* New York: Peter Lang.

———. (2006a). New labour's education policy. In D. Kassem, E. Mufti, and J. Robinson (eds.). *Education Studies: Issues and Critical Perspectives.* Buckingham: Open University Press.

———. (2006b). Class, capital and education in this neoliberal/neoconservative period. *Information for Social Change*, 23. Online at http://libr.org/isc/issues/ISC23/B1%20Dave%20Hill.pdf

Hill, D. (2007). Critical teacher education, new labour in Britain, and the global project of neoliberal capital. *Policy Futures*, 5(2), 204–225. Online at http://www.wwwords.co.uk/pfie/content/pdfs/5/issue5_2.asp.

Hill, D. (ed.). (2009a). *The Rich World and the Impoverishment of Education: Diminishing Democracy, Equity and Workers' Rights.* New York: Routledge.

Hill, D. (ed.). (2009b). *Contesting Neoliberal Education: Public Resistance and Collective Advance.* London, New York: Routledge.

Hill, D., and Kumar, R. (eds.). (2009). *Global Neoliberalism and Education and Its Consequences.* New York: Routledge.

Hill, D., and Rosskam, E. (eds.). (2009). *The Developing World and State Education: Neoliberal Depredation and Egalitarian Alternatives.* New York: Routledge.

Hill, D. with Anijar-Appleton, K., Davidson-Harden, A., Fawcett, B., Gabbard, D., Gindin, J., Kuehn, L., Lewis, C., Mukhtar, A., Pardinaz-Solis, R., Quiros, B., Schugurensky, D., Smaller, H., and Templer, B. (2006). Education services liberalization. In E. Rosskam (ed.). *Winners or Losers? Liberalizing Public Services.* Geneva: International Labour Organisation.

Hill, D., Maisuria, A., and Greaves, N. (2006). Does capitalism inevitably increase education inequality? In D.B. Holsinger and W.J. Jacob (eds.). *International Handbook on Educational Inequality.* Dordrecht, The Netherlands: Springer.

Hirtt, N. (2004). Three axes of merchandisation. *European Educational Research Journal,* 3(2), 442–453. Available at http://www.wwwords.co.UK/eerj/.

———. (2005) The growth of high-stakes testing in the U.S.: accountability, markets and the decline in educational equality. *British Education Research Journal,* 31(5), 605–622.

International Chamber of Commerce (ICC). (1999). *The Benefits of Services Trade Liberalisation: Policy Statement.* Document 103/210. Paris: ICC. Available at: http://www.iccwbo.org/home/case_for_the_global_economy/benefits_services_tr ade_liberalization.asp.

Johnson, P., and Lynch, F. (2004). Sponging off the poor. *The Guardian,* March.10.

Jones, K., Cunchillos, C., Hatcher, R., and Hirtt, N. (2007). *Schooling in Western Europe: The New Order and Its Adversaries.* London: Palgrave MacMillan. Online at http://www.amazon.co.uk/Schooling-Western-Europe-Order-Adversaries/dp/0230551432/ref=sr_1_7?ie=UTF8&s=books&qid=1254948680&sr=8-7

Korten, D. (2004). Article in *EDUcate.* Sindh Educational Foundation, 3(2).

Leher, R. (2004). A new lord of Education? World Bank policy for peripheral capitalism. *Journal for Critical Education Policy Studies,* 2(1). Available at http://www.jceps.com/index.php?pageID=article&articleID=20.

Mahoney, P., I. Menter, and I. Hextall. (2003). Edu-business: are teachers working in a new world? *Paper given at American Education Research Association Annual Conference,* Chicago, April 21–25. Available at http://www.roehampton.ac.uk/perc/iptatw/AERAEdu-business.pdf.

Marx, K., and Engels, F. (1848) [1950]. The Communist Manifesto. In *Karl Marx and Frederick Engels, Selected Works.* London: Lawrence & Wishart.

McLaren, P. (2000). *Che Guevara, Paolo Freire and the Pedagogy of Revolution.* Lanham, ML and Oxford: Rowman and Littlefield.

McLaren, P. (2005) *Capitalists and Conquerors: A Critical Pedagogy Against Empire.* Lanham, MD: Rowman and Littlefield.

McLaren, P. (ed.). (2006). *Rage and Hope: Interviews with Peter McLaren on War, Imperialism and Critical Pedagogy.* New York, Peter Lang.

McLaren P., and Farahmandpur, R. (2005). *Teaching against Global Capitalism and the New Imperialism.* Lanham, MD: Rowman & Littlefield.

McLaren, P., and Rikowski, G. (2001). Pedagogy for revolution against education for capital: an e-dialogue on education in capitalism today. *Cultural Logic: An Electronic Journal of Marxist Theory and Practice,* (4) no. 1, 1. Available at http://eserver.org/clogic/4–1/mclaren&rikowski.html and at http://www.ieps.org.uk.cwc.net/mclarrikow.pdf.

Molnar, A. (2001). *Giving Kids the Business: The Commercialisation of America's Schools* (second edition). Westview: Harper Collins.

———. (2005). *School Commercialism: From Democratic Ideal to Market Commodity.* London: RoutledgeFalmer.

Molnar, A., Wilson G., and Allen, D. (2004). *Profiles of For-Profit Education Management Companies, Sixth Annual Report, 2003–2004.* Arizona: Arizona State University Education Policy Studies Laboratory. Available at http://www.asu.edu/educ/epsl/CERU/Documents/EPSL-0402–101-CERU.pdf.

Office for National Statistics. (2000). *Living in Britain 2000.* Online at http://www.statistics.gov.UK.

Pollin, R. (2003). *Contours of Descent.* London: Verso.

Rikowski, G. (2002). Globalisation and education. A paper prepared for the House of Lords Select Committee on Economic Affairs, Report on "Globalisation," HL Paper 5–1, November 18. On House of Lords CD-ROM. Also at *Independent News for Student and School student Activists.* Available at http://education.portal.dk3.com/article.php?sid=21 and at www.ieps.org.uk.

———. (2003). Schools and the GATS enigma. *Journal for Critical Education Policy Studies,* 1(1). Available at http://www.jceps.com/index.php?pageID=article&articleID=8.

———. (2005a). *Silence on the Wolves: What is Absent in New Labour's Five Year Strategy for Education.* Occasional Paper published by the Education Research Centre, University of Brighton, UK. *Falmer Papers in Education,* 1(1).

———. (2005b). The education white paper and the marketisation and capitalisation of the secondary schools system in England. October 24, in Two Parts: Part I is at: http://journals.aol.co.uk/rikowskigr/Volumizer/entries/571 and Part II at: http://journals.aol.co.uk/rikowskigr/Volumizer/entries/572.

Ross, E.W., and Gibson, R. (eds.). *Neoliberalism and Education Reform.* Cresskill, NJ: Hampton Press.

Rosskam, E. (ed.). (2006). *Winners or Losers? Liberalizing Public Services.* Geneva: International Labour Organisation.

Santos, B. (2004). *A universidade no século XXI.* Sao Paolo, Brazil: Cortez.

Schugarensky, D., and Davidson-Harden, A. (2003). From Cordoba to Washington: WTO/GATS and Latin American education. *Globalization, Societies and Education,* 1(3).

Sinclair S., and Grieshaber-Otto J. (2002). *Facing the Facts: A Guide to the GATS Debate.* Ottawa: Canadian Centre for Policy Alternatives.

Siqueira, A. (2005). The regulation of education through the WTO/GATS: path to the enhancement of human freedom? *Journal for Critical Education Policy Studies,* 3(1).

Weiner, J. (2006) UCLA's "Dirty Thirty." *The Nation,* February 13. Available at http://www.thenation.com/doc/20060213/wiener.

Whitty, G., Power, S., and Halpin, D. (1998). *Devolution and Choice in Education: The School, the State and the Market.* Buckingham, UK: Open University Press.

Chapter 6

Defending Dialectics: Rethinking the Neo-Marxist Turn in Critical Education Theory

Wayne Au

Introduction

Of all of Marx's extensive writing on political economy and philosophy, there is perhaps no other single segment that has produced as much theoretical wrangling, conflict, and discussion about the relationship between capitalism and social structures as that contained in the "Preface to a Contribution to the Critique of Political Economy," where Marx (1968b) writes:

> In the social production of their life, [humans] enter into definite relations that are indispensable and independent of their will, relations of production which correspond to a definite stage of development of their material productive forces. The sum total of these relations of production constitutes the economic structure of society, the real foundation, on which rises a legal and political superstructure and to which correspond definite forms of social consciousness. The mode of production of material life conditions the social, political and intellectual life process in general. It is not the consciousness of [humans] that determines their being, but, one the contrary, their social being that determines their consciousness. (P. 183)

These four sentences outline what is commonly referred to as the base/superstructure model in Marxism.

It has been commonplace to critique Marx's earlier mentioned conception for placing too much emphasis on the economy as the singular determinant of social relations and sociopolitical institutions. In the field of education this has translated into a criticism of Marxism as positing a direct, linear, mechanical, or functionalist correspondence between the needs of the capitalist economy and the structures of schooling. In response to such critiques, critical educational theorists have placed more emphasis on the "superstructure" of society and culture through a focus on the concepts of "relative autonomy" of the superstructure from the economic base (Althusser, 1971) and "hegemony" (Gramsci, 1971). These added distinctions presumably distinguish neo-Marxism from what is now termed "traditional" or "orthodox" Marxism.

The purpose of this chapter is to explore the foundations of this debate. First, I discuss the application of economic determinism within critical educational theory and outline the substantial, and correct, critiques of this analysis by neo-Marxists. Then, I outline how critical educational theorists turned to the work of Althusser and Gramsci in the interest of developing neo-Marxist theory that explains inequality in education but still allows for the agency of individuals and relative autonomy of education in relation to capitalist production. Finally, I return to the works of Marx and Engels to explore their conception of the relationship between society and capitalist production, and find the turn to neo-Marxism as unwarranted, given the nondeterministic, nonfunctionalist nature of Marx and Engel's original formulation.

Education and the Correspondence Principle

Contemporary arguments surrounding the relationship between schools and the reproduction of capitalist economic relations were largely sparked by Bowles and Gintis' *Schooling in Capitalist America: Educational Reform and the Contradictions of Economic Life* (1976). Bowles and Gintis advance the "correspondence principle" of educational relations, where in capitalist societies,

> the division of labor in education, as well as its structure of authority and reward, mirror those of the economy ... [and] in any stable society in which a formal educational system has a major role in the personal development of working people, there will tend to emerge a correspondence between the social relations of education and those of the economic system. (Bowles and Gintis, 1988, p. 237)

Bowles and Gintis' formulation asserts that schools simply function to serve the needs of capitalist production in nearly a one-to-one correspondence, and offers a mechanistic interpretation of Marx's analysis of the relationship between the economic base and superstructure, quoted earlier.

Critical education theorists correctly criticized Bowles and Gintis' correspondence principle, arguing that it ignores the role of teachers, culture, and ideology in schools, is too mechanical and overly economistic, and neglects students' and others' resistance to dominant social relations (see, e.g., Apple, 1979/2004, 1995; Carlson, 1988; Cole, 1988; Giroux, 1980; Moore, 1988; Sarup, 1978; Sharp, 1980). Faced with the mechanical formulation of a functionalist, one-to-one correspondence between the relations of production (economic base) and their reproduction in schools (the superstructure), some critical educational theorists were compelled to develop a critique of capitalist relations in schools and society while simultaneously allowing for resistance to those dominant relations. By conflating Marxism with the economic determinism of scholars such as Bowles and Gintis, these theorists turned to the works of Althusser and Gramsci for a solution to this dilemma, thus establishing the neo-Marxist tradition of analysis within critical educational theory. In what follows I provide an explanation of these theorists' concepts and how they have been used by neo-Marxist educational theorists.

Althusser and Relative Autonomy

Althusser, a French communist and philosopher, is another theorist whose ideas have been widely adopted by neo-Marxist educational scholars, largely due to his application of the concept of relative autonomy. In his discussion of the relationship between the economic base (what he refers to as the "infrastructure") and the superstructure, Althusser (1971) states:

> It is easy to see that this representation of the structure of every society as an edifice containing a base (infrastructure) on which are erected the two "floors" of the superstructure, is a metaphor, to be quite precise, a spatial metaphor...this metaphor...suggests that the upper floors could not "stay up" (in the air) alone, if they did not rest precisely on their base...Thus the object of the metaphor of the edifice is to represent above all the "determination in the last instance" by the economic base. (P. 135)

Given the superstructure's "determination in the last instance" by the base, Althusser arrives at two conclusions: "(1) there is a 'relative autonomy' of the superstructure with respect to the base; (2) there is a 'reciprocal action' of the superstructure on the base" (p. 136).[1] Additionally, he identifies two distinct components of the superstructure: the Repressive State Apparatus (RSA) and the Ideological State Apparatus (ISA), which the powerful use to maintain hegemonic control through force and ideology, respectively.

For Althusser it is the ISA, of which schools are a part, which maintains ideological hegemony for the ruling class. Indeed, Althusser sees state-sanctioned education as central in maintaining hegemony. For instance, in his discussion of the role of schools he states, "I believe that the ideological State apparatus which has been installed in the *dominant* position in mature capitalist social formations…is the *educational ideological apparatus*" (p. 153; emphasis in the original). Further, he goes on to discuss how schools, as a tool of bourgeois hegemony, are presented as a universally neutral and natural mechanism (p. 157). Althusser's conception of schools in relation to hegemony meshes with Gramscian view in that schools transmit the "universally reigning ideology" while simultaneously maintaining the image of being a "neutral environment purged of ideology." Thus, they contribute to, in Gramscian terms, the "spontaneous consent" of the dominated.

It is Althusser's concept of relative autonomy that has been widely adopted by neo-Marxist educational theorists. For instance, Apple (1995) explains that:

> [T]here was as dynamic interplay between the political and economic spheres which was found in education. While the former was not reducible to the latter—and, like culture, it had a significant degree of relative autonomy—the role the school plays *as a state apparatus* is strongly related to the core problems of accumulation and legitimation faced by the state and a mode of production…(Emphasis in the original; p. 26)

Strands of Althusser's formulation can also be found running through the work of theorists such as Bourdieu and Passeron (1977), Bernstein (1990), and Apple (2002). Althusser's concept of relative autonomy is also used within what is loosely referred to "resistance theory" in education. For instance, Giroux (1983) asserts that: "In resistance accounts, schools are relatively autonomous institutions that not only provide spaces for oppositional behavior and teaching but also represent a source of contradictions that sometimes make them dysfunctional to the material and ideological interests of the dominant society" (p. 260). Resistance theory, as evidence of the relative autonomy of schools, takes on issues of cultural production and reproduction as central fields of inquiry, and is perhaps epitomized by the ethnographic work of Willis (1977, 2003), who found that working class British school boys "resisted the mental and bodily inculcations of the school and rejected students who showed conformist attitudes to the school authorities" (Willis, 2003, pp. 392–393). The concept of relative autonomy thus holds a utilitarian value for resolving the problems posed by economic determinism and aids critical education theorists in developing theories of resistance (Dance, 2002; Giroux, 1981, 1983; Willis, 1977, 2003), because it attempts to both acknowledge human intervention through cultural practices and to establish schools as spaces where the possibility of social transformation might be created.[2]

Althusser's conception is contradictory, however. Ironically, while he does challenge economic determinism broadly through the concept of relative autonomy, he is also noted for denying human subjectivity and agency in relation to social, economic, and historical structures. This has been termed Althusser's "antihumanism" where: "[T]he self, the human subject, does not so much constitute but is constituted by the structural, systemic relations in which it finds itself. It is the belief not that [humans] make history but that history makes [humans] or that history makes itself..." (Smith, 1985, p. 649). Indeed, in *For Marx*, Althusser (1969) argues that Marx's early analysis of human agency should be thrown out in favor of an analysis that focuses solely on structural influence (Poster, 1974). Critiques of Althusser's emphasis on structuralism also appear in critical educational theory, where his work is even grouped with the economic determinism of Bowles and Gintis (see, e.g., Giroux, 1980, 1983, 2003) because it belies the power of humans, as agentic subjects, to act in the world.

Gramsci and the Concept of Hegemony

Antonio Gramsci, a founder of the Italian Communist Party, was imprisoned by the Italian Fascists in 1926. While in prison he penned over three thousand pages of notebooks on Marxist theory and political strategies for his party (Allman, 1999; Coben, 1995), subsequently edited and published in English translation as *Selections From The Prison Notebooks* (Gramsci, 1971).[3] Gramsci's exploration and elaboration of the concept of *hegemony* has become a central tenet in neo-Marxist theorizing. Based on a close textual reading of *The Prison Notebooks*, Allman (1999) explains:

> [Gramsci] uses the term "hegemony," or moral, ethical leadership, to describe the means by which consent is organized. However, hegemony

> is a form of leadership that can work primarily by either domination or direction (i.e., leading). In his analysis of how hegemony works in bourgeois civil society, he describes how it works primarily by domination or imposition of ideological systems of belief as well as through the absorption of radical elements into the existing framework. (Pp. 105–106)

Gramsci's concept of hegemony is intimately linked to his formulation of the relationship between the superstructure and the economic base. He specifies two ways in which the superstructure reproduces capitalist relations: The first is hegemonic—through ideology and universalized "spontaneous consent"—while the second is through "legal" enforcement of judiciaries and other institutions associated with the state (Gramsci, 1971, p. 12). With his focus on the processes of hegemony and domination and the state's role in the two, Gramsci places an emphasis on the superstructure's relative autonomy from the economic base (Carnoy, 1982).

Gramsci (1971), however, does not elevate the superstructure to independent status. Rather, he conceives of the superstructure as being dialectically related to the economic base. As he states:

> [Economic] structures and superstructures form an "historical bloc." That is to say the complex, contradictory and discordant *ensemble* of the superstructure is the reflection of the *ensemble* of the social relations of production. From this, one can conclude: that only a totalitarian system of ideologies gives a rational reflection of the contradiction of the structure . . . This reasoning is based on the necessary reciprocity between structure and superstructure, a reciprocity which is nothing other than the real dialectical process. (P. 366; original emphasis)

Gramsci's description is of a superstructure that is dynamic, fluid, and consisting of an assemblage of parts, suggesting a dialectical, nonfunctionalist conception of the relationship, a conception that, as I shall address later, is not a significant departure from Marx's original analysis.

In placing a focus on the superstructure, identifying its two levels, expanding the concept of hegemony, and asserting the partial autonomy of the state, thus Gramsci provided neo-Marxists with a way to address the dilemma presented by mechanical, functionalist analyses of the relationship between the economic base and the social/educational superstructure. For instance, understanding hegemony as a process in and of itself that takes place partially through education allows it to be explicitly targeted, analyzed, and potentially disrupted—a tactical position that has been correctly noted by critical educational theorists (see, e.g., Apple, 1980). The position that education provides a site for resistance to bourgeois hegemony is a key concept for neo-Marxist theorists. Gramsci's emphasis on the potential autonomy of the state/superstructure in relation to the economic base places schools in a position of not being fully determined by the economic structure and reinforces their potential for resisting bourgeois hegemony (Allman, 1999, 2001; Apple, 1979/2004, 1995, 2002; Coben, 1995; Giroux, 1983, 1999).

Additionally, Gramsci's conception of hegemony has also been used within critical educational theory to explain how consent of the subordinate is essentially "won" by those in power. Often, to maintain their legitimacy, dominant elites offer compromises or accords with subordinate groups (Apple and Buras, 2006), accords that can act as "an umbrella under which many groups can stand but which basically still is under the guiding principles of dominant groups" (Apple, 2000, p. 64). This particular application of hegemony has been used within critical educational policy studies, for instance, to frame and understand why some segments of racially oppressed communities lend their support to the movement to privatize public education via the use of vouchers—a movement that ultimately increases social inequality (Apple, 2006; Pedroni, 2007). Such an analysis finds that the conservative Right uses appeals to racial equity to gain the support of communities of color, who see their immediate interests served through the implementation of a policy such as school vouchers. Thus, Gramsci's conception of hegemony also allows for critical education theorists to recognize some amount of human agentic action as individuals and communities that consciously interact with social, economic, cultural, political, and educational "structures."

Thus far I have explained how the functionalist, economic determinist explanation of the relationship between schools and capitalist production was correctly deemed inadequate by critical educational theorists. Further, I have explained how these theorists turned toward Althusser's conception of relative autonomy and Gramsci's concept of hegemony in search of a less restrictive analysis that could still maintain a Marxist critique of capitalism while allowing for human agency and consciousness—thus giving rise to neo-Marxist critical education theory. What I have found in this historical-theoretical study, however, is a problematic within neo-Marxist theorizing. Mainly, much of the neo-Marxist critique has erroneously conflated functionalist, economic determinism with Marxism. In this sense, we can see the disservice of Bowles and Gintis' (1976) analysis: Their functionalist, economic determinist account epitomizes what today is often referred to as "orthodox Marxism" almost as if it were common sense (see, e.g., Leonardo, 2004; Smith, 1984).[4] However, as Apple (1982) correctly observes, "Marx himself consistently employed the ideas of base and superstructure in a complex way. Rather than calling for an economistic perspective where 'the economy' produces everything else, we find a much more substantive usage" (p. 10). Marx and Engels themselves have a tradition of struggling against economistic and mechanical interpretations of Marxist theory, and, as we shall see, their conception is complex enough to raise doubts about the necessity of the turn toward "neo"-Marxism.

Marx and Engels' More Substantive Usage

In their texts, Marx and Engels offer a conception of the relationship between the economic base and the superstructure that is dynamic and nonfunctionalist. For instance, although they did not take up an explicit study of the concept, the roots of a complex analysis of hegemony exist in "The German Ideology" (Marx and Engels, 1978), where they state:

> The ideas of the ruling class are in every epoch the ruling ideas: i.e., the class which is the ruling *material* force of society, is at the same time its ruling *intellectual* force... The ruling ideas are nothing more than the ideal expression of the dominant material relationships, the dominant material relationships grasped as ideas;... The individuals composing the ruling class... rule as a class and determine the extent and compass of the epoch, it is self-evident that they do this in its whole range, hence among other things rule also as thinkers, as producers of ideas, and regulate the production and distribution of the ideas on their age: thus their ideas are the ruling ideas of the epoch... (Pp. 172–173; emphasis in the original)

They go on to add that:

> For each new class which puts itself in the place of one ruling before it, is compelled, merely in order to carry through its aim, to represent its interest as the common interest of all the members of society, that is, expressed in ideal form: it has to give its ideas the form of universality, and represent them as the only rational, universally valid ones... Every new class, therefore, achieves its hegemony only on a broader basis than that of the class ruling previously... (P. 174)

These offerings of Marx and Engels are open to a complex analysis of hegemony that does several things. First, it begins to interrogate the relationship between ideology and power in society in a way that recognizes how those in control have the power and capacity to produce and distribute their ideas, and that this power and capacity rests on their relative control over material production. Further, Marx and Engels' analysis includes the concept of ideological universality, where the interests of the ruling elite are presented as the common interests of the whole society. Marx also addresses the concept of hegemony in "The Class Struggles in France, 1848–1850" (1978) where he raises the role of inter-class and intra-class conflict and compromise in the social, political, and economic turmoil in France at the time. Thus, while Gramsci may have considered hegemony as a specific concept more deeply in his notebooks, Marx and Engels express what might be considered a proto-conception of hegemony that is congruent with Gramsci's later work (Carnoy, 1982).

In concert with developing a proto-conception of hegemony, Marx and Engels also provide a nuanced and complex analysis of the relationship between the economic base and the superstructure. In a letter to J. Bloch, Engels (1968d) critiques economistic interpretations as gutting Marxist conceptions of history:

> According to the materialist conception of history, the *ultimately* determining element in history is the production and reproduction of real life. More than this neither Marx nor I have ever asserted. Hence if somebody twists this into saying that the economic element is the *only* determining one, he transforms that proposition into a meaningless, abstract, senseless phrase. The economic situation is the basis, but the various elements of the superstructure—political forms of the class struggle and its results, to wit: constitutions,...judicial forms,...political, juristic, philosophical theories, religious views and their further development into systems of dogmas—also exercise their influence upon the course of the historical struggles and in many cases preponderate in determining their *form*. There is an interaction of all these elements in which, amid all the endless host of accidents, the economic movement finally asserts itself as necessary...(P. 692; emphasis in the original)

In this passage Engels clarifies the Marxist conception of the base/superstructure relationship. After establishing that he and Marx never asserted that economics was the sole determining factor, and recognizing that the superstructure does play a role in shaping history, Engels adds that the superstructure, "in many cases preponderate in determining [the] *form*" (emphasis in the original) of class struggle.

This point in particular speaks to issues of resistance, human agency, and mediation of bourgeois hegemony within the superstructure, including schools. In essence, it posits a type of relative autonomy to the superstructural elements that Engels outlines, thus opening the door for a Marxist analysis that asserts that various superstructural elements, human consciousness and human action "exercise their influence upon the course of historical struggles" even as the "economic movement finally asserts itself as necessary" (Engels, 1968d, p. 692; see also Engels, 1968e).

Further, Engels (1968c), in a letter to H. Borgius, addresses the role of capitalist economic necessity in relation to the superstructure:

> Political, juridicial, philosophical, religious, literary, artistic, etc., development is based on economic development. But all these react upon one another and also upon the economic basis. It is not that the economic situation is *cause, solely active*, while everything else is only passive effect. There is, rather, interaction on the basis of economic necessity, which *ultimately* always asserts itself... So it is not, as people try here and there conveniently try to imagine, that the economic situation produces an automatic effect. No. [Humans] make their history themselves, only they do so in a given environment, which conditions it, and on the basis of actual relations already existing, among which the economic relations... are still ultimately the decisive ones. (Pp. 704–705; emphasis in the original)

It is the economic necessity, the movement created by the contradictions inherent in capitalist production, that "ultimately always asserts itself" over the superstructure. It is in this dialectical sense that the economic basis is the driving force for the superstructure "in the last analysis" (Gramsci, 1971, p. 162) or "in the last instance" (Althusser, 1971, p. 135).

As we see from the evidence presented earlier, Marx and Engels explained the relationship between the economic base and the superstructure in terms of relational interaction, not in terms of economic functionalism. This does not mean, however, that Marx and Engels denied that the state could at times have relative autonomy from the economic base. On the contrary, Marx and Engels explained the relative autonomy of the state both historically and theoretically. For instance, in "The Eighteenth Brumaire of Louis Bonaparte," Marx (1968a) offers historical evidence of the relative autonomy of the state. He observes that "Only under the second Bonaparte does the state seem to have made itself completely independent... And yet the state power is not suspended in mid air. Bonaparte represents a class..." (p. 171). Engels (1968c) offers a more theoretical explanation of the relative autonomy when he states that:

> The further the particular sphere which we are investigating is removed from the economic sphere and approaches that of pure abstract ideology, the more shall we find it exhibiting accidents in its development, the more will its curve zigzag. But if you plot the average axis of the curve, you will find that this axis will run more and more nearly parallel to the axis of economic history. (P. 705)

Engels' point here is particularly provocative. Not only is he advocating that other "spheres" can behave somewhat autonomously in relation to the "economic sphere," but he posits that we can only qualify the economic base and superstructure *relationally* with each other. Further he asserts that the more autonomous the state or aspects of the superstructure are from the economic base, the more they contradict the needs of the relations of production. This distance is expressed relatively through "accidents" in the development of a "particular sphere."

However, it is apparent that, in stressing the power that the economic production wields over the superstructure, many of Marx and Engels' contemporaries interpreted their analysis as being economic determinist. For instance, Engels (1968b), in a letter to F. Mehring, laments:

> Marx and I always failed to stress enough in our writings in regard to which we are all equally guilty. That is to say, we all laid, and *were bound to lay*, the main emphasis, in the first place, on the *derivation* of political, juridicial and other ideological notions, and of actions arising through the medium of these notions, from basic economic facts. But in so doing we neglected the formal side—the ways and means by which these notions, etc., come about—for the sake of content. This has given our adversaries a welcome opportunity for misunderstandings and distortions...(P. 700; emphasis in the original)

In an explanation of why he and Marx emphasized the economic base in relation to the superstructure, Engels (1968d) says that:

> Marx and I are ourselves partly to blame for the fact that the younger people sometimes lay more stress on the economic side than is due to it. We had to emphasise the main principle *vis-à-vis* our adversaries, who denied it, and we had not always the time, the place or the opportunity to give their due to the other elements involved in the interaction. (P. 693)

Further, in a critique of mechanistic formulations of Marxist theory by their contemporaries, Engels (1968b) explains:

> It is above all this semblance of an independent history of state constitutions, of systems of law, of ideological conceptions in every separate domain that dazzles most people... This aspect of the matter, which I can only indicate here, we have all, I think, neglected more than it deserves. It is the old story: form is always neglected at first for content... Hanging together with this is the fatuous notion of the ideologists that because we deny an independent historical development to the various ideological spheres which play a part in history we also deny them any *effect upon history*. The basis of this is the common undialectical conception of cause and effect. (P. 701; emphasis in the original)

Marx and Engels were so focused on developing a materialist understanding of capital and capitalism that Engels admits they neglected issues of ideological and state form—partially as an expression of their human physical limitations. (In spite of the monumental amount of work he produced, Marx simply died before he was able to finish all the volumes of capital.) In the same stroke, Engels does correctly place blame for the critique of Marxism for not allowing for enough "independent history of state constitutions, of systems of law, or ideological conceptions" squarely on an "undialectical conception of cause and effect." As I have argued here, such undialectical conceptions led neo-Marxists to critique Marxism for economic determinism in the first place.

Defending Dialectics

Based on the above evidence, the neo-Marxist conflation of Marxism with functionalist, economic determinism within critical educational theory needs to be problematized. If we take dialectical materialism as the core philosophical perspective of Marxist theory,[5] then it is a logical impossibility to equate an "orthodox" Marxist account (at least in the sense of Marx and Engels' original dialectical materialist conception) with a functionalist, economic determinist account of social reproduction in education (Sayers, 1990). This is because dialectical relationships are dynamic, interactional, fluid, and relational and therefore do not allow for linear, mechanical, one-to-one chains of causality or correspondence (Allman, 1999; Engels, 1940; Ollman, 2003; Sayers, 1990; Woods and Grant, 2002). This is the type of analysis we see in Engels' explanation of the relationship between the economy and society, discussed earlier in several instances. It is a relationship characterized by "zigzags," "the wills of individuals," "exhibiting accidents," and "interactions" between various aspects of the superstructure and the base of capitalist economic production. It is apparent that Engels in fact disagreed with his contemporaries who interpreted Marx's analysis to mean that there was a direct, automatic cause and effect relationship between the economy and society. As Engels argues of these particular critics of Marxism:

> What these [people] all lack is dialectics. They always see only here cause, there effect. That this is a hollow abstraction, that such metaphysical polar opposites exist in the real world only during crises, while the whole vast process goes on in the form of interaction—though of very unequal forces, the economic movement being by far the strongest, most primordial, most decisive—that here everything is relative and nothing absolute—this they never begin to see...(Engels, 1968a, p. 699)

To posit a functionalist, deterministic, one-to-one correspondence between schools and capitalist production is decidedly antidialectical, or "undialectical" in Engels' words, and therefore arguably does not fall within the tradition of Marxist analysis.

The question still remains: What is the relationship between schools and the reproduction of capitalist social relations? Some advocate that the base/superstructure metaphor within Marxist, critical educational analysis be completely rejected (see, e.g., Rikowski, 1997). It is not clear to me, however, that the problem is with the metaphor itself. Rather, based on the arguments presented here, the central problem seems to have been maintaining a dialectical materialist analysis when we think about the relationships and processes at hand. It is when their analyses slip into the linear, mechanical logics associated with the rationalist tradition (Benton and Craib, 2001; Sayers, 1990) that critical theorists generally run into difficulty with Marxism.

As Sayers (1990) points out, one of the key tenets of Marxist dialectics is that in order for us to understand something as it concretely exists as part of material reality, "[I]t is vital to see them in the context of their interconnections with other things within a wider whole" (p. 143). Arguably, this has been the intent of all Marxist, neo-Marxist, and functionalist formulations of the relationship between schools and society—the empirical evidence connecting schools and inequality is too overwhelming to deny such a relationship. The devil, however, is in the details of *how* we conceive of the interconnections between things within an organic, interrelated totality. Again, Sayers (1990) provides a glimpse of such a conception in his summation of Marxist dialectical relations:

> Social processes have their own internal dynamic, their own inner contradictions. The different aspects of society—forces and relations of production, base and superstructure—are aspects of a single whole, internally and organically interrelated, in dialectical interaction and conflict. It is these interactions, these conflicts, these contradictions—which are internal to society—that lead to historical change. In the process, none of these aspects is inert or passive: the forces and relations of production and also the superstructure are all transformed and developed. (P. 164)

The importance of understanding social and economic processes as having their own internal dynamics cannot be overstated, for it recognizes that there are logics of development at play within these relationships, that social and economic systems in a sense have their own life *and* are simultaneously made up of the lives of individual humans. As Creaven (2000) observes,

> The existence of such relationships of structural dependence (of polity, law, major cultural institutions, etc.,) upon economic production and exploitation is what justifies the Marxist view that societies are systems, or totalities, following their own logics of development, rather than a heterogeneous ensemble of "autonomous" structures or practices, moving in no particular direction. (P. 67)

The conception of these relationships as systematic, as processes that develop in particular directions and that exhibit particular characteristics, function in ways that can be interrogated, understood, and, ultimately, changed.

Thus, in a Marxist conception, schools, as part of the superstructure, have a contradictory relationship with the relations of capitalist production. As Fritzell (1987) explains regarding the contradictory nature of the state's relationship with the economic base:

> [It] could be argued that in a functional context the autonomy of the State refers essentially to a *potentiality*, insofar as it is granted that even under empirical conditions of advanced capitalism the State cannot in the long run enforce policies and interventions that are basically destructive to the commodity form of economic production. (Emphasis in the original; p. 27)

Fritzell roots the essential contradiction of the position of the state in the fact that it is fundamentally outside of the process of producing commodities, yet it still is required under capitalism to support the production of those commodities. In relation to capitalist production and social reproduction, the state is thus required to work out this internal contradiction. Apple (1995) provides a corollary analysis of schools, as part of the state, when he observes:

> On the one hand, the school must assist in accumulation by producing both agents for a hierarchical labor market and the cultural capital of technical/administrative knowledge. On the other hand, our educational institutions must legitimate ideologies of equality and class mobility, and make themselves be seen as positively by as many classes and class segments as possible. In a time of fiscal crisis, the structural contradiction is exacerbated. The need for *economic* and ideological efficiency and stable production tends to be in conflict with other *political* needs. What we see is the school attempting to resolve what may be the inherently contradictory roles it must play. (P. 53; emphasis in original)

In this statement, Apple gets at the root of the relationship between schools and social reproduction. Schools, on behalf of the state-superstructure, have to simultaneously accomplish the fundamentally contradictory goals of reproducing the social and material relations of capitalist production while hegemonically working to win the "spontaneous consent" of the students/workers through appeals to individual equality within the educational and social meritocracy. This contradiction presents a dialectical relationship between production of capitalist social relations and the maintenance of bourgeois hegemony vis-à-vis education.

The important piece of the neo-Marxist impetus, however, is to recognize that within a Marxist, dialectical analysis, human beings are not totally determined beings. As Marx (1968a) himself asserted in the oft-quoted "The Eighteenth Brumaire of Louis Bonaparte," humans "make their own history, but they do not make it as they please; they do not make it under self-selected circumstances, but under circumstances existing already, given and transmitted from the past" (p. 97). Or, in Engel's (1968e) words, "In the history of society . . . the actors are all endowed with consciousness, are [humans] acting with deliberation or passion, working towards definite goals; nothing happens without a conscious purposes, without an intended aim" (p. 622). Indeed, within a Marxist conception, humans do have agency, they can be and are subjects of history. This was the goal of both Lenin's (1975) and Vygotsky's (1978, 1987) conceptions of consciousness (Au, 2007) and is the backbone of Freire's (1974) conception of "liberatory pedagogy" (Au and Apple, 2007): that humans, as subjects, as agents, as individuals, and as individual classes, develop consciousness of the imposition of structures on their lives and, based on that consciousness, take action to change those structures. Total determination and subjugation simply cannot exist within a Marxist, dialectical materialist conception.

Further, as Anderson (1980, chapter 2) remarks, the terms "agent" and "subject" both are internally contradictory: agent signifies both "active initiator" and "passive instrument" (e.g., the agent of a foreign power); and subject signifies both "sovereignty" and "subordination." Such internal contradiction perhaps points to the appropriateness of both terms, for it provides analytic space, in a Marxist conception, for both individual consciousness and schools to be "relatively autonomous" from the relations of production associated with the economic base. Thus, while schools play a key role in reproducing social inequality, their contradictory role in legitimating ideologies of equality also allows room for resistance to this reproduction (Apple, 1979/2004, 1995; Carnoy and Levin, 1985). It is absolutely crucial for us to recognize this room for resistance because students *do* resist the inculcations of schooling on many levels (Au, 2005; Dance, 2002; Shor, 1992; Willis, 1977, 2003), and teachers, as laborers within the political economy of education (Apple, 1986, 1995), also resist the reproduction of inequitable capitalist socialist relations in their classrooms and schools (Allman, 1999; Allman, McLaren, and Rikowski, 2000; Carlson, 1988; Freire, 1974; Shor, 1992). In this way, a dialectical conception of the relationship between schools and capitalism, in a Marxist, dialectical materialist sense, poses a significant challenge to the economic determinism of Bowles and Gintis, one that still recognizes that the superstructure is emergent from, but not reducible to, the economic base (Apple, 2000; Creaven, 2000).

Conclusion

In this chapter I have argued that the scholarly debate amongst critical educational theorists regarding the relationship between schools and the reproduction of dominant social relations associated with capitalism hinges on an erroneous conflation of Marxism with functionalist, economic determinist analyses, and such a conflation represents the disappearance of dialectics from critical educational theory. Further, I have outlined how many critical educational theorists thus turned to the concepts of hegemony and relative autonomy in order to address the shortcomings of functionalist analyses by incorporating ideas of individual agency, resistance, cultural expression, and subjectivity that they felt Marxism lacked. Last, this chapter finds that Marx and Engels recognized a nonlinear, dynamic, nonmechanistic, *dialectical* relationship between the economic base and the superstructure—one that acknowledges that the superstructure and the state have an effect on history and the forms in which class struggle and bourgeois hegemony are expressed through the complex, chaotic social relations of capitalist production. Such a conception is a far cry from the critique of functionalist, economic determinism often laid at the feet of Marxism by neo-Marxists and anti-Marxists alike.

I have sought here to constructively critique the neo-Marxist rejection of Marxist orthodoxy, for this rejection is based on an incorrect conflation of Marxism with economic determinism. In the process I have performed, in Marxist, dialectical materialist terms, what would be considered a negation of the negation (Ollman, 2003): Where neo-Marxism sought to negate the perceived economic functionalism of Marx and Engels' original conceptions, the analysis provided here has in turn negated that negation, sublating, or subsuming, the concepts of hegemony and relative autonomy within a more refined understanding of Marxist dialectics. Thus we are left with a Marxist conception of schools and society that does indeed account for resistance to the pressures of capitalist production within schooling and education, but one that still acknowledges the defining role that capitalist production plays in the outcomes of our systems of education.

Notes

1. It is not a very far stretch to map Gramsci's formulation of "civil society" and "political society," what he identifies as the two categorical parts of the state in his later writings, onto Althusser's "Ideological State Apparatus" (ISA) and "Repressive State Apparatus" (RSA), respectively.
2. I do not fully endorse Althusser's conception. Indeed, in my own analysis of "relative autonomy," I arrive at the conclusion that Althusser provides for too much autonomy to the superstructure, and therefore opens up the space for analyses of culture, the superstructure and/or the state that are divorced from the material reality of capitalist production. This raises the question as to whether or not Althusser's conception was indeed philosophically materialist, a prerequisite for Marxist analysis. For a critique of the use of relative autonomy and resistance theories in education, see Rikowski (1997).

3. It should be noted that various aspects of Gramsci's ideas have been claimed by a wide-ranging and often contradictory set of political perspectives, including educational conservatives (i.e., E.D. Hirsch), Marxists, Leninists, post-Marxists, neo-Marxists, feminists, post-structuralists, and postmodernists (Buras, 1999; Giroux, 1999; Holst, 1999; Jessop, 2001). Jessop (2001) qualifies Gramsci's theorizing as being of "incomplete and tentative character," thus making it "compatible with several other theoretical currents" (p. 151). Indeed, Gramsci had to write cryptically in order to avoid the censors, and he died before getting the chance to edit the notebooks himself, allowing for increased confusion on the meanings of key terms such as ideology, state, and hegemony (Allman, 1999; Giroux, 1983). The editors and translators of *Selections from the Prison Notebooks* attribute this confusion partially to Gramsci's own mixed usage of terms and partly to issues related to translation into English (see Hoare and Smith in the preface to Gramsci, 1971, pp. xii–xiv)
4. My critique of Leonardo here must be seen as friendly, for I very much respect and appreciate both Leonardo and his work in critical educational theory.
5. I recognize that this is a position that some "analytic Marxists," such as Cohen (1978) and Wright (1997), challenge. For a critique of analytic Marxism, see Sayers (1990) and Roberts (1996).

References

Allman, P. (1999). *Revolutionary Social Transformation: Democratic Hopes, Political Possibilities and Critical Education* (first edition). Westport, CT: Bergin & Garvey.

———. (2001). *Critical Education against Global Capitalism: Karl Marx and Revolutionary Critical Education* (first edition). Westport, CT: Bergin & Garvey.

Allman, P., McLaren, P., and Rikowski, G. (2000). After the box people: the labour-capital relation as class constitution—and its consequences for Marxist educational theory and human resistance. Retrieved May 1, 2004, from http://www.ieps.org.uk.cwc.net/afterthebox.pdf.

Althusser, L. (1969). *For Marx*. New York: Pantheon Books.

———. (1971). *Lenin and Philosophy and Other Essays* (trans. B. Brewster). New York: Monthly Review Books.

Anderson, P. (1980). *Arguments within English Marxism*. London: Nlb.

Apple, M.W. (1979/2004). *Ideology and Curriculum* (third edition). New York: RoutledgeFalmer.

———. (1980). "The other side of the hidden curriculum: Correspondence theories and the labor process." *Interchange*, 11(3), September.

———. (1986). *Teachers and Texts: A Political Economy of Class and Gender Relations in Education*. New York: Routledge & Kegan Paul.

Apple, M.W. (1995). *Education and Power* (second edition). New York: Routledge.

———. (2000). *Official Knowledge: Democratic Education in a Conservative Age* (second edition). New York: Routledge.

———. (2002). Does education have independent power? Bernstein and the question of relative autonomy. *British Journal of Sociology of Education*, 23(4), 607–616.

———. (2006). *Educating the "Right" Way: Markets, Standards, God, and Inequality* (second edition). New York: Routledge.

Apple, M.W. (Ed.). (1982). *Cultural and Economic Reproduction in Education: Essays on Class, Ideology and the State* (first edition). Boston: Routledge & Kegan Paul.

Apple, M.W., and Buras, K.L. (Eds.). (2006). *The subaltern Speak: Curriculum, Power, and Educational Struggles.* New York: Routledge.

Au, W. (2005). Power, identity, and the third rail. In P.C. Miller (ed.). *Narratives from the Classroom: An Introduction to Teaching* (pp. 65–85). Thousand Oaks, California: Sage Publications.

———. (2007). "*Vygotsky and Lenin: parallel structures of individual and social development.*" Science & Society, 71(3), July, 273–298.

Au, W., and Apple, M.W. (2007). Freire, critical education, and the environmental crisis. *Educational Policy*, 21(3), 457–470.

Benton, T., and Craib, I. (2001). *Philosophy of Social Science: The Philosophical Foundations of Social Thought.* Houndmills, Basingstoke, Hampshire; New York: Palgrave.

Bernstein, B.B. (1990). *The Structuring of Pedagogic Discourse* (first edition, vol. IV). New York: Routledge.

Bourdieu, P., and Passeron, J. (1977). *Reproduction in Education, Society, and Culture.* Beverly Hills, CA: Sage.

Bowles, S., and Gintis, H. (1976). *Schooling in Capitalist America: Educational Reform and the Contradictions of Economic Life* (first edition). New York: Basic Books.

———. (1988). Schooling in capitalist America: reply to our critics. In M. Cole (ed.). *Bowles and Gintis Revisted: Correspondence and Contradiction in Educational Theory* (first edition). Philadelphia: The Falmer Press.

Buras, K.L. (1999). Questioning core assumptions: a critical reading of and response to E.D. Hirsch's *the schools we need and why we don't have them. Harvard Educational Review*, 69(1), 67–93.

Carlson, D.L. (1988). Beyond the reproductive theory of teaching. In M. Cole (ed.). *Bowles and Gintis Revisited: Correspondence and Contradiction in Educational Theory* (pp. 158–173). New York: The Falmer Press.

Carnoy, M. (1982). Education, economy, and the State. In M.W. Apple (ed.). *Cultural and Economic Reproduction in Education: Essays on Class, Ideology and the State* (pp. 79–126). Boston: Routledge & Kegan Paul.

Carnoy, M., and Levin, H.M. (1985). *Schooling and Work in the Democratic State.* Stanford, California: Stanford University Press.

Coben, D. (1995). Revisiting Gramsci. *Studies in the Education of Adults*, 27(1), 36–52.

Cohen, G.A. (1978). *Karl Marx's Theory of History: A Defense* (expanded edition). Princeton: Princeton University Press.

Cole, M. (Ed.). (1988). *Bowles and Gintis Revisted: Correspondence and Contradiction in Educational Theory* (first edition). Philadelphia: The Falmer Press.

Creaven, S. (2000). *Marxism and Realism: A Materialistic Application of Realism in the Social Sciences.* London; New York: Routledge.

Dance, J.L. (2002). *Tough Fronts: The Impact of Street Culture on Schooling* (first edition). New York: RoutledgeFarmer.

Engels, F. (1940). *Dialectics of Nature* (trans. C. Dutt; first edition). New York: International Publishers.

———. (1968a). Engels to C. Schmidt in Berlin. In *Karl Marx & Frederick Engels: Their Selected Works* (pp. 694–699). New York: International Publishers.

———. (1968b). Engels to F. Mehring in Berlin. In *Karl Marx & Frederick Engels: Their Selected Works* (pp. 699–703). New York: International Publishers.

———. (1968c). Engels to H. Borgius in Breslau. In *Karl Marx & Frederick Engels: Their Selected Works* (pp. 704–706). New York: International Publishers.

———. (1968d). Engels to J. Bloch in Konigsberg. In *Karl Marx & Frederick Engels: Their Selected Works* (pp. 692–693). New York: International Publishers.

———. (1968e). Ludwig Feuerbach and the end of classical German philosophy. In I. Publishers (ed.). *Karl Marx & Frederick Engels Selected Works* (pp. 596–618). New York: International Publishers.

Freire, P. (1974). *Pedagogy of the Oppressed* (trans. M.B. Ramos). New York: Seabury Press.

Fritzell, C. (1987). On the concept of relative autonomy in educational theory. *British Journal of Sociology of Education*, 8(1), 23–35.

Giroux, H.A. (1980). Beyond the correspondence theory: notes on the dynamics of educational reproduction and transformation. *Curriculum Inquiry*, 10(3), 225–247.

———. (1981). Hegemony, resistance, and the paradox of educational reform. *Interchange*, 12(2–3), 3–26.

———. (1983). Theories of reproduction and resistance in the new sociology of education: a critical analysis. *Harvard Educational Review*, 53(3), 257–293.

———. (1999). Rethinking cultural politics and radical pedagogy in the work of Antonio Gramsci. *Educational Theory*, 49(1), 1–19.

———. (2003). Public pedagogy and the politics of resistance: notes on a critical theory of educational struggle. *Educational Philosophy and Theory*, 35(1), 5–16.

Gramsci, A. (1971). *Selections from the prison notebooks* (trans. Q. Hoare and G.N. Smith). New York: International Publishers.

Holst, J.D. (1999). The affinities of Lenin and Gramsci: implications for radical adult education theory and practice. *International Journal of Lifelong Education*, 18(5), 407–421.

Jessop, B. (2001). Bringing the State back in (yet again): reviews, revisions, rejections, and redirections. *International Review of Sociology*, 11(2), 149–173.

Lenin, V. I. (1975). *What is to be Done?: Burning Questions of our Movement.* Peking: Foreign Language Press.

Leonardo, Z. (2004). The unhappy marriage between Marxism and race critique: political economy and the production of racialized knowledge. *Policy Futures in Education*, 2(3 & 4), 483–493.

Marx, K. (1968a). The eighteenth brumaire of Louis Bonaparte. In *Karl Marx & Frederick Engels: Their Selected Works* (pp. 95–180). New York: International Publishers.

———. (1968b). Preface to a contribution to the critique of political economy. In *Karl Marx & Frederick Engels: Their Selected Works* (pp. 181–185). New York: International Publishers.

———. (1978). The class struggles in France, 1848–1850. In R.C. Tucker (ed.). *The Marx-Engels Reader* (second edition; pp. 586–593). New York: W.W. Norton and Company Ltd.

Marx, K., and Engels, F. (1978). The German ideology: part I. In R.C. Tucker (ed.). *The Marx-Engels Reader* (pp. 146–200). New York: W.W. Norton & Company.

Moore, R. (1988). The correspondence principle and the Marxist sociology of education. In M. Cole (ed.). *Bowles and Gintis Revisited: Correspondence and Contradiction in Educational Theory* (pp. 51–85). New York: The Falmer Press.

Ollman, B. (2003). *Dance of the Dialectic: Steps in Marx's Method* (first edition). Chicago: University of Illinois Press.

Pedroni, T. C. (2007). *Market Movements: African American Involvement in School Voucher Reform.* New York: Routledge.

Poster, M. (1974). Althusser on history without man. *Political Theory*, 2(4), 393–409.

Rikowski, G. (1997). Scorched earth: rebuilding Marxist educational theory. *British Journal of Sociology of Education*, 18(4), 551–574.

Roberts, M. (1996). *Analytical Marxism: A Critique.* London; New York: Verso.

Sarup, M. (1978). *Marxism and Education.* Boston: Routledge & Kegan Paul.

Sayers, S. (1990). Marxism and the dialectical method: a critique of G.A. Cohen. In S. Sayers and P. Osborne (eds.). *Socialism, Feminism, and Philosophy: A Radical Philosophy Reader* (pp. 140–168). New York: Routledge.

Sharp, R. (1980). *Knowledge, Ideology, and the Politics of Schooling: Towards a Marxist Analysis of Education.* London; Boston: Routledge & Kegan Paul.

Shor, I. (1992). *Empowering Education: Critical Teaching for Social Change* (first edition). Chicago: The University of Chicago Press.

Smith, S. B. (1984). Althusser and the overdetermined self. *The Review of Politics*, 46(4), 516–538.

———. (1985). Althusser's Marxism without a knowing subject. *The American Political Science Review*, 79(3), 641–655.

Vygotsky, L.S. (1978). *Mind in Society*. Cambridge, MA: Harvard University Press.

———. (1987). Thinking and speech (trans. N. Minick). In R.W. Rieber and A. Carton (eds.). *The Collected Works of L.S. Vygotsky: Problems of General Psychology Including the Volume Thinking and Speech* (vol. 1; pp. 37–285). New York: Plenum Press.

Willis, P. (1977). *Learning to Labor: How Working Class Kids Get Working Class Jobs*. New York: Columbia University Press.

———. (2003). Foot soldiers of modernity: the dialectics of cultural consumption and the 21st-century school. *Harvard Educational Review*, 73(3), 390–415.

Woods, A., and Grant, T. (2002). *Reason in Revolt: Dialectical Philosophy and Modern Science* (North American Edition; vol. 1). New York: Algora Publishing.

Wright, E.O. (1997). *Class Counts: Comparative Studies in Class Analysis*. New York: Cambridge University Press.

Chapter 7

Hijacking Public Schooling: The Epicenter of Neo-Radical Centrism

João M. Paraskeva

A Neo-Radical Centrism

Understanding and analyzing neoliberal globalization involves an accurate set of critical hermeneutical processes that dig extensively into the very marrow of the cultural, economic, and political origins of these policies (Sousa Santos, 2005). Neoliberal globalization—in its multiplicity of forms—did (and is) not happen(ing) in a social vacuum. Actually, "it is precisely in its oppression of non-market forces that we see how neoliberalism operates not only as an economic system, but as a political and cultural system as well" (McChesney, 1999, p. 7; Olssen, 2004), a factor that creates endless intricate tensions between cultural homogenization and cultural heterogenization (Appadurai, 1996). Thus, accurately examining the forms of neoliberal globalization implies a cautious consideration of the emergence of Reaganism-Bushism and Thatcherism-Majorism in the United States and England, respectively (Sousa Santos, 2005). These ideologies and their associated policies were responsible for the origins of a cultural revolution that, among other issues, initiated a feverish and frenetic attack, not simply on the Welfare State and its apparatuses, but on the very idea behind it, highlighting the market not only as a solution but as the only one. The 1980s will be known as the "Reagan decade" (House, 1998, p. 18), or as a period that witnessed a conservative "right turn" that renounced the commonsense meanings of the particular central social concepts that underpin a just society (Hall, 1988). Such a "right(ist) turn" needs to be understood as a non-monolithic bloc that has been able to edify an intricate and powerful coalition incorporating seemingly antagonistic groups—neoliberals, neoconservatives, authoritarian populists, and a fraction of a new middle class (Apple, 2000). As I have documented elsewhere[1] (where I have conceptualized and justified Apple's organic intellectuality as anchored in three trilogies), Apple's outline might not be a proper fit to explain particular frightening realities in Zimbabwe, Rwanda, Angola, or Mozambique. However, it helps us understand that, as Sousa Santos (2005) claims, "the neoliberalism contrary to what is commonly maintained, is not a new form of liberalism, but rather a new form of conservatism" (p. vii), in which discretely specific yet powerful religious groups are steadily assuming prominent power positions. In fact, the present role played by O(cto)pus Dei, a Vatican within the Vatican (Hutchison, 2006) and a kind of sophisticated expression of light Christi-fascism (Kincheloe and Steinberg, 2007), threatens the very way we examine the hegemonic forces behind neoliberal globalization and their strategies.

In analyzing the latest metamorphosis of *New Rightist* policies, Mouffe (2000, p. 108) stresses that both Blair and Clinton were able to construct a "radical center." Unlike traditional political groupings, the radical center is a new coalition that "transcends the traditional left/right division by articulating themes and values from both sides in a new synthesis" (p. 108). However, as I have examined elsewhere,[2] unlike Mouffe, Fairclough (2000) stresses that the radical center strategy does not consist only in "bringing together elements from these [left and right] political discourses" but also in its ability to "reconcile themes which have been seen as irreconcilable beyond such contrary themes and transcending them" (pp. 44–45). Fairclough (ibid.) also argues that this strategy is not based on a dialogic stance. That is to say, the radical center achieved consent within the governed sphere "not through political [democratic] dialogue, but through managerial methods of promotion and forms of consultation with the public; [that is to say] the government tends to act like a corporation treating the public as its consumers rather than its citizens" (p. 129). While such radical centrism targets the state, Hill (2003) claims that neoliberal forces actually need a strong state to promote their interests, especially in areas such as education and training—fields that are deeply related to the formation of an ideologically submissive labor force. What has evolved then is a state that has fostered the development of "the magnet economy" (Brown and Lauder, 2006) so that "whatever the market cannot provide for itself, the state must provide for it" (Gabbard, 2003, p. 65). It is actually the state that has been paving the way for the market (Gabbard, 2003; Macrine, 2003; Paraskeva, 2003, 2004, 2008; Sommers, 2000).

As Appadurai (1996) and Olssen (2004) claim, although from different angles, state sovereignty has never been in jeopardy within the contemporary global cultural flows. In essence, neoliberal imprimatur is a result of nonstop struggles between state and market forces. Such intricate tensions are the needed fuel for the neoliberal intellectual engines (Paraskeva, 2001, 2008). In fact, such radical centrism, while searching for the dissolution of old contradictions between "right" and "left" (Ferguson, 2001), was able to lay a solid foundation for the gradual emergence of a new concept of state (especially with regard to its role), anchored in a need to modernize government at almost any cost. Democratic forces have been colonized by managerial insights in such a way that governments end up being weak executives of a *Res* plc[3] that operates with the blessing of an anemic popular vote (Ferguson, 2001).

We argue that such mercantilistic neo-fundamentalism has paved the way for what Agamben (2005) called a "State of exception"—the embryo of what I have called neo-radical centrism. While radical centrism claims to offer a broad managerial concept for the public good by showing new managerial dynamics per se (Newman, 2001, p. 46), *neo*-radical centrism actually refines the entire commonsense cartography edified and sutured by radical centrism. What is at stake nowadays for the neo-radical centrists—profoundly influenced and driven by menacing O(cto)pus Dei judicious desires—is not the rapacious need for modernizing forms of governments but rather the unbalanced tension between force and law. In short, force transcends law, paradoxically, in so-called democratic nations.

The issue today is so complex that it goes well beyond Clarke and Newman's (1997) brilliant analysis found in *The Managerial State* in which they challenge, among other things, the tension(s) of welfare without a state. Currently the issue goes well beyond, as Clarke, Gewirtz and McLaughlin (2001) accurately argue, the creation of mixed economies of welfare, or the emergence of a new public management, the transformation of citizens into consumers, or even the emergence of forms of entrepreneurial government.

The issue today is that the state is stumbling before the tyranny of transformation in a way already flagged by Clarke and Newman (1997). Ironically, Welfarecide was originally orchestrated and supported by so-called radical centrist policies, but now neo-radical centrism emerges as an answer to the compound framework of needs resulting as a consequence of just such Welfarecide. Thus, while radical centrism cannot be seen as a crises but an answer to the crises (Apple, 2000), neo-radical centrism cannot be seen as simply a need but rather as only an answer to address ever more pressing needs. As Agamben (2005, p. 1) argues—anchored in Schmitt's (1922) approach—"the necessities transcend the law." In this way, Agamben (2005, p. 1) is able to overcome the multifarious tensions prompted by "state of exception *vs.* state sovereignty" and edifies a "point of imbalance between public law and political fact." In fact, Agamben claims that the state of exception "appears as the legal form of what cannot have legal form." Neo-radical centrism is "ambiguous, uncertain, borderline fringe, at the intersection of the legal and the political" (pp. 1–2) in its layout, making it conveniently well-situated in coded no man's land and quite juicy for marketers.

If we view recent events such as *oilgate, foodgate, biodieselgate* as the dangerous costs of democratic mumble, we will be able to critically recognize how the state of exception "tends increasingly to appear as the dominant paradigm of government in contemporary politics, [a] threshold of indeterminacy between democracy and absolutism" (pp. 2–3).

Based on a need to defend the values that support an insolent, Western, eugenic culture, the state of exception (which is a state of necessities) has legitimized the conditions for the naturalization and legalization of torture and genocide, allowing mass imprisonments, massive extermination of particular categories of (supposedly evil) individuals, and the fabrication of new identities. For instance, it has become commonplace to talk about the "detainees in Guantanamo Bay" and not about "POWs" who should be treated according to the Geneva Convention. Simultaneously, the State of Exception reinforces the conditions that anchor societal development to a pale economic equation. In fact, "the state of exception is not a special kind of law (like the law of war)...[Quite conversely] it is a suspension (in our understanding *ad eternum*) of the juridical order itself" (p. 4; Todorov, 2003). In essence, such neo-radical centrism wins the consensus, with the vast majority of the population assuming neofundamentalistic perspectives (p. 20). Todorov (p. 21) argues that we are not witnessing the rise of a conservative, neoconservative, or even paleo-conservative movement, but rather that of a "neofundamentalist hegemonic bloc." Odd as it might be, Todorov writes that such a political platform gathers such disparate political entities as Maoists and Trotskyists. Unfortunately, some of the best Marxists are on the right center (Paraskeva, 2006).

Why then, despite almost three decades of distressing effects on society and attacks on "the even more localized rest" (Bauman, 2004, p. 3), does a hegemonic bloc continue to dominate? As Jessop et al.

(1984) and Apple (2000, p. 23) remind us, one must question, "How is such an ideological vision legitimated and accepted?" It is undeniable that neo-radical centrism is not exactly a pure detour from the orthodoxies laid out by radical centrism. It is actually a moment of complexities, and in some ways it is a platform that, as Hall (1992) would put it, leans toward radical centrism by taking advantage of particular kind of contradictions within the very marrow of neoliberal globalization. Neo-radical centrism should be seen as the latest capitalist metamorphosis of righting the left. Such an aim cannot be detached from the politics of the commonsense and the role that the media plays in building a particular weave. Notwithstanding its disastrous impact on oppressed populations all over the world, such a new hegemonic power bloc managed to achieve support from that majority on the social perimeter.

The Struggle over the Commonsense

Until recently, the concepts of equality, freedom, and democracy informed the sociopolitical role of public schooling in producing, promoting, and suturing the common good. Sadly, these ideals have been gradually perverted, twisted, and, in a sense, almost deleted from the vocabulary of daily life. The purpose of public education has been gradually transformed. Such a transformation process achieves a superior level of complexity under the neo-radical centrist platform. In a way, public schools have been emptied of any sense of public good: a critical scheme within the strategy of highjacking the public schools.

Concepts are not inert entities, and their meanings are constructed within a particular context. We therefore need to pay attention to the "meaning of language in its specific context" (Apple, 2000, p. 16). In so doing, we become capable of "understanding political conceptions and educational concepts, since they are part of a larger social context," a context that is "constantly shifting and is subject to severe ideological conflicts" (ibid.). Education "itself is an arena in which these ideological conflicts work themselves out, [since] it is one of the major sites in which different groups with distinct political, economic, and cultural visions attempt to define what socially legitimate means and ends of a society are to be" (p. 17; Giroux, 1981; Macedo, 2006). Apple (2000) denounces the shifting meanings of the word "equality" "that have a good deal of success in redefining what education is for and in shifting the ideological texture of the society profoundly to the right" (p. 17).

In this rather refined, truncated, and truculent process of redefinition of the very meaning and purpose of public schooling, we see an unquestioned fundamentalist faith in the ability of free choice to address social and educational chaos, combined with attacks on teachers and curriculum on issues such as quality, accountability, commitment, high-stakes testing. It is, as Bartolome (2007), Gillborn (1990, 2004), Hursh (2008), Lipman (1998), Valenzuela (1999), and others well documented, a faith that "ignores" how such policies have been profoundly coded by class and race (and one could add gender as well). Essentially, the extraordinary political strategy led by the neoliberals drove both the social and educational worlds into a set of multifarious conflicts, conflicts that have led to a substantive transformation of schooling to the right (Ball, 2005; Macedo, 2006; Whitty, 2002). As Apple (2000) highlights, "one of the major aims of a rightist restoration politics is to struggle in not one but many different arenas at the same time, not only in the economic sphere but in education and elsewhere as well" (p. 21). Thus, in order to succeed in this endeavor, "the economic dominance must be coupled to 'political, moral, and intellectual leadership" (ibid.).

Using this intricate strategy that acts dynamically not only within the economic sphere but also in other societal sites as well, the neo-hegemonic bloc successfully interceded within the common-sense environment, one that is deeply ordinary and contradictory, interrupting, renovating, and transforming the very consciousness of people. It is precisely within this judicious restructuring of the commonsense (a complex outcome of multifaceted and contradictory accords) that cultural battles are fought. Indeed, like radical centrism, neo-radical centrism doesn't give up its capacity to work and rework the common sense and to generate new meanings among key societal concepts.

The New Right is transforming and distorting the meanings of key words, pushing basic concepts such as public versus private, democracy, autonomy, diversity, freedom, social justice, equality, and human rights into the economic realm where they have gradually lost their capacity to address and build the public common good (Paraskeva, 2003, 2004). As Macedo, Dendrinos, and Gounari (2003) argue, "in order to redefine the concept of freedom, neoliberal ideology produces a powerful discourse whose effects are so pervasive that it becomes almost impossible for anybody to even imagine freedom outside the market order" (p. 124). The reworking of the meanings of particular key words to gradually reconfigure the commonsense to one that serves the purpose of the New Right agenda implies a careful and intricate process of (dis)articulation and (re)articulation. Such a process, Torfing (1999) claims, allows one to understand "how cultural artifacts are overdetermined by political ideologies, and by social and political identities in terms of class, race, nationality, and gender" (p. 211). Hence, articulation, as Hall (1996) reminds us, is the "form of the connection that 'can' make a unity of two different elements, under certain conditions; [that is to say] it is a linkage which is not necessary, determined, absolute, and essential for all time" (p. 141). "While under the radical center platform the State was reduced to a pale entity and the government tends to act like a corporation treating the public as its consumers rather than citizens" (Fairclough, 2000, p. 12). In other words under the neo-radical centrist insights we were faced with a neofundamentalist (yet discrete) restate of the state, which is, as we have been arguing, a state of exception. The configuration of key societal meanings has been complicated from one platform to another. The very process of articulation (Laclau and Mouffe, 1985, p. 105) has become much more sophisticated.

As Klein (2007) describes the Katrina tragedy, Friedman (*Uncle Miltie*) did not blink when defending the view of such a tragedy as a great opportunity to radically reform the educational public system by giving vouchers to affected families so that they could be used in state-funded private institutions. Another example was unveiled in Macrine's (2003) analysis of the public school crisis in a Philadelphia school district, in which "the state had plans to sell off the management of schools to the highest bidder." U.S. Republican presidential candidate John McCain pushed and promoted this same philosophy at several Republican rallies.

Having analyzed the way the commonsense is reconstructed and reorganized around the tension of old words, it is necessary to underscore how the mainstream media has overtly lent a hand in this intricate and smooth process. This recasting of language helps to perpetuate a particular hegemonic commonsense that maintains and refines the constant stage of upgrade of the capitalist project, a project that needs to be situated in what Arrighi (1994) calls the third hegemonic period within the history of capitalism. As Macedo (2006) perfectly stresses, it is "through the manipulation of language that the ideological doctrinal system is able to falsify and distort reality, making it possible for individuals to accommodate to life within a lie" (p. 39). Neo-radical centrism has been able to win the struggle over specific and crucial sets of meanings. On these particular issues the media do play a key role.

The media is constrained by both political and economic censorship. In fact, the promiscuity between media apparatuses and government institutions within a capitalist system is an undisputable reality. As McChesney (2004) argues, our media, "far from being on the sidelines of the capitalist system, are among its greatest beneficiaries" (p. 21). We are before a particular form of ideological control that is profoundly related to what Bourdieu (1996) calls a "show and hide" strategy. In essence, one should be aware that in the tension between "giving news *vs.* giving views" (p. 42), the mainstream media aligns with the latter.

One can say that under the free market economy, the media acts according to what Herman and Chomsky (2002, p. 1) and Chomsky (2002b) call the "propaganda mode [and] it is their function to amuse, entertain, and inform, and to inculcate individuals with the values, beliefs, and codes of behavior that will integrate them into the institutional structures of the larger society." Mainstream media act dynamically and overtly in the construction of a complex amalgam of what Fairclough (1992) coined as "social identities," "subject positions," "self typologies," and "systems of knowledge and beliefs." McCain's campaign rally is but one example of the way the media highlights fighting, supporting belligerent U.S. eugenic perspectives.

We are under a distorted construction of the "other," and simultaneously a twisted fabrication of what I called in another context (Paraskeva, 2005, p. 268) an "endemic western we," a " 'we' that consistently define[s] international events, especially foreign social movements, in ways that [confirm] the dominant political meanings and values of the Western white hegemony" (2005, p. 268). In a way *unavoidability* is a cultural terrain daily prepared by mainstream media apparatuses flooded with a multiplicity of meanings that have been edifying particular kinds of realities and laying out the pillars for specific future *inevitable happenings.* Iraq, Afghanistan, and Iran are but three good examples of this claim.

The media edifies particular kinds of ideological filters (Chomsky, 2002a) that maintain a nonstop process of building specific "communities of interpretation," communities that are built through constant struggle yet only within particular semantic borders that reward a specific kind of commonsense (Said, 1997, pp. 36–68).

It is precisely this "propaganda model" that actively promotes what Bourdieu (1996) calls "the circular circulation of information" (p. 23). Given rating dynamics and the race for audiences, the media is competing over the same issues, and "in some sense, the choices made on television are choices made by no subject" (p. 25). Therefore, as one can easily see, "television puts 'reality' together" (Eldridge, 1993, p. 4; Paraskeva, 2007b). In so doing, as Derrida (2002) argues, television "produces an artifactuality" (p. 42). The news is not something that is "out there." Quite conversely, news is something "doable" in a very biased way.

It is because of this pertinent concern that one can identify with Molnar's (1996a,b) Apple's (2000), McLaren and Farahmandpour's (2002), and Macrine's (2003) analysis of the effects of Channel One within schooling through both the news and the commercials. Channel One should be seen as a "paradigm case of the social transformation of our ideas about public and private, and about schooling itself" (Apple, 2000, p. 111). Channel One is a clear evidence of the way the mainstream media "plays" in the complex process of reconfiguring the commonsense through the intricate process of articulation (Fiske, 1986; Paraskeva, 2007b). We are confronted with a political, economic, and cultural scheme to construct a "captive audience" and a daily praxis that is "selling our kids to business" (Molnar, 1996a,b)—despite some noteworthy forms of resistance, as Macrine (2003) pointedly reminds us. In fact, a poem *I am Not For Sale,* written by "a high school student and member of the Philadelphia student union named Sara" (pp. 210–211), is clear evidence not only of the difficulties that the neoliberal(ism) bloc is facing, but also and above all is an example of a tenacious and elaborated form of resistance led by students to challenge dehumanizing policies.

This issue becomes even more complex when one considers advertising and how it interplays with identity formation. The role of advertising in shaping desires and opinions was discussed by Williams (1974), to whom "the sponsorship of programmes by advertisers has an effect beyond the separable announcement and recommendation of a brand name" (p. 68).

One of the major arguments against Channel One is its insidious covert construction of semi-intransitive consciousness (Freire, 1974). Underneath the creation of these semi-intransitive consciousness lies an insidious process of identity construction. Thus, it would be a "mistake to conceive of the concept of the audience as only an economic category [since] the concept of the audience is intricately bound up with the dimensions of social and cultural identity" (Grossberg, Wartella, and Whitney, 1998, p. 207). Thus, the audience is indeed a social construction (re)producing identities.

In fact, the construction of captive audiences is a very powerful strategy of "hailing the subject by [particular kinds] of discourse"

(Hall, 1988, p.6). The mainstream media does play a key role in the struggle over meanings and in so doing helps create a distorted and perverted view of the so-called third world, such as the "natural" floods in South America and Southern Africa, the Students' Revolt in Paris, the political situation in Zimbabwe, or the ongoing genocide in Sudan, Ruanda, and the shameful reality of African immigration in the South of Spain. Mainstream media has a consistent record of defending eugenic policies and practices to defend and perpetuate what Steinberg and Kincheloe (2001) call the "white supremacist power bloc" (p. 17)

Despite the extent of its pernicious reach, neo-radical centrism is experiencing unexpected obstacles, such as, say, democracy and the state. As Saramago (2003) claims, like in ancient Rome, "the main and definite obstacle to the implementation of democracy [nowadays comes] precisely from the power of an economic aristocracy that saw the democratic system as a real direct enemy of their interests" (p. 8). Undeniably, this is the Achilles' heel of contemporary Western democracies, a weakness that opens the door for the concubinage relations between the state and the market (Saramago, 2003; Chomsky, 1989). Saramago (2003) adds, "people did not elect their government for the government to drive the people to the market. The market uses all of its capacity to condition the government so the government takes the people to them" (pp. 8–9).

Moreover, Saramago highlights the uselessness of a "democracy that it is not built on the basis of a truly economic democracy and of an effective and powerful cultural democracy" (p. 9). As he claims "don't even bother assuming the tremendous responsibility of killing democracy, since 'democracy' is committing suicide every single day" (ibid.). It is in this context that Negri (2002) refuses to reinvent democracy arguing, "if you force me to reinvent democracy I simply cannot do it. I am tired of a system that perfectly adapts itself to capitalism" (p. 175).

Saramago's (2003) approach allows one to perceive that a real democratic system that embraces economic, cultural, and political justice and equity is a powerful obstacle to neo-rightist purposes. Such an approach opens the door to an accurate understanding of something very crucial to education. In fact, given the relations between the state and the market, the real and effective democratic meaning and practices of common good defended by a myriad of societal spheres dissipates gradually. One of those spheres is public schooling—a prime target of the new-rightist strategy. Public schooling has been kidnapped and is on the verge of having its final blessing in many nations. Such strategy has been paved by a wise set of chirurgical defunding policies (Apple, 2003; Ball, 2001; Hursh, 2006; Macedo and Bartolomé, 2001; Mouffe, 2006; Paraskeva, 2007a,b; Paraskeva, Ross, and Hursh, 2006; Torres Santomé, 2001; Whitty and Power, 2002). Such dangerous disinvestments functioned as the needed sign for the market forces to kidnap public schools and convert schooling into one of the market's guinea pigs. As the Portuguese neo-radical centrists claim, "the quasi monopoly of public schooling is not desirable (...) since it contradicts the constitutional principle of freedom to teach and to learn, to choose a social good" (XV Governo Constitucional de Portugal, 2001). Weiner's (2005) analysis between Althusser's Ideological State Apparatuses (ISA) and Bauman's Ideological Corporate Apparatuses (ICA) suggests that, despite their differences, both the ideological and state apparatuses aim to "reproduce the conditions of reproduction through pedagogical strategies (...) to make reality a matter of commonsense" as well (p. 23).

Another Knowledge is Possible: Toward Reinventing Social Emancipation

A very good way to challenge neo-radical centrist policies and strategies and to fight for a more just society is to engage in a struggle that respects what Sousa Santos, Nunes, and Meneses (2007) call epistemological diversity. As they argue, there is no such thing as "global social justice without cognitive justice" (p. ix). In fact, by claiming as "official" particular kinds and forms of knowledge, schooling does participate in what Sousa Santos (1997) called espistemicides—a lethal tool that fosters the commitment to imperialism and white supremacy (hooks, 1994). In fact, Sousa Santos, Nunes, and Meneses (2007) claim quite astutely that the "suppression of knowledge [from indigenous peoples of the Americas and of the African slaves] was the other side of genocide" (p. ix).

Thus, one cannot deny that "there is an epistemological foundation to the capitalist and imperial order that the global north has been imposing to the global south" (ibid.). As Sousa Santos (2004) argues, we need to engage in a battle against "the monoculture of scientific knowledge" and fight for an "ecology of knowledges," which is "an invitation to the promotion of non-relativistic dialogues among knowledges, granting equality of opportunities to the different kinds of knowledge" (Sousa Santos, Nunes, and Meneses, 2007, p. xx). The target of our fighting should be the coloniality of power and knowledge. In so doing, we will end up challenging the notions, concepts, and practices of multiculturalism that are profoundly

> Eurocentric, created and describ[ing] cultural diversity within the framework of the nation-states of the Northern hemisphere (...) the prime expression of the cultural logic of multinational or global capitalism, a capitalism without homeland at last, and a new form of racism, tend[ing] to be quite descriptive and apolitical thus suppressing the problem of power relations, exploitation, inequality, and exclusion. (Pp., xx–xxi)

In so doing we will be allowing for fruitful conditions, or what Sousa Santos (2004) calls the sociology of absences that will challenge invisible subjectivities (Ellison, 1952).

What we have here is a call for the democratization of knowledge, a commitment toward an emancipatory non-relativistic and cosmopolitan knowledge. We need to learn from the South (since) the aim to reinvent social emancipation goes beyond the critical theory produced in the North and the social and political praxis to which it has subscribed (Sousa Santos, Nunes, and Meneses, 2007, p. xiv).

This is a herculean task, but one that cannot be denied if one is committed to realizing a just society. Actually, as Cox (2002) reminds us, "globalization is [also] a struggle over knowledge of world affairs" (p. 76). The struggle against the Western eugenic coloniality of knowledge will transform schools and their social agents into real leaders in the tough endeavor to democratize democracy. As Sen (1999, p. 3) claims, the emergence of democracy was the event of the twentieth century. For him, the real question is not "Is a whether a country was "fit for democracy," but rather the prevailing view in the twentieth century is that a country has to become "fit through democracy."

Sousa Santos, (2005) argues that our current time is marked by huge developments and dramatic changes, an era coined as the electronic revolution of communications and information, genetic and biotechnological revolution. On the other hand,

> it is a time of disquieting regressions, a return of the social evils that appeared to have been or about to overcome, [a] return of slavery and slavish work [a] return of high vulnerability to old sicknesses that seemed to have been eradicated and appear now linked to new pandemics like HIV/AIDS [a] return of the revolting social inequalities. (P. vii)

It is the role of teachers as public intellectuals, Giroux (1994) argues, to decenter the curriculum, in a way to separate it from its westerning forms and content, to fight savage inequalities (Kozol, 1999). Schooling has to play a leading and key role in addressing one of the most challenging issues that we have before us—democratizing democracy. It is undeniable, Vavi (2004) claims, that democracy is bypassing the poor.

Sousa Santos (2005) is quite accurate when he claims that we are living in an era with modern problems without modern solutions. In order to re-democratize democracy, Sousa Santos (2005) suggests that we need to reinvent social emancipation since traditional social emancipation has been pushed into a kind of dead end by neoliberal globalization.

However, a different form of globalization, a counter-hegemonic globalization that will propel a myriad of social movements and transformations, has challenged such globalization. It is exactly within the very marrow of such counter-hegemonic forms of globalization and in its clashes with the neoliberal hegemonic globalization that new itineraries of social emancipation are developing.

Thus, the struggle for democracy, Shivji (2003) argues, "is primarily a political struggle on the form of governance, thus involving the reconstitution of the state and creating conditions for the emancipatory project" (p. 1). Somehow, we are clearly before what Sousa Santos (1998) coined as a state that should be seen as a spotless new social movement. That is, a more vast political organization in which the democratic forces will struggle for a distributive democracy, thus transforming the state into a new—yet powerful—social and political entity. The task is to daily reinvent how to democratize democracy. We need, Nussbaum (1997) argues,

> to foster a democracy that is reflective and deliberative, rather than simply a market place of competing interest groups, a democracy that genuinely takes thought for the common good [...] it is not good for democracy when people vote on the basis of sentiments they have absorbed from talk radio and never questioned. (P. 19)

A new struggle must, in fact, begin. This is the best way, as the, Mozambican writer Couto (2005) claims, that we have "to challenge a past that was portrayed in a deformed way, a present dressed with borrowed clothes and a future ordered already by foreign interests" (p. 10). As Nyerere wisely claims (1998) it will be judicious not to choose money as our weapon since, "the development of a country is brought about by people, not by money" (p. 129)—something that, painfully, marketers seem to neglect. Public education does have a key role in claiming that (an)other knowledge is possible and how such knowledge is crucial for the transformative processes of democratizing democracy. As Aronowitz (2001)—one of Horowitz's anecdotally listed one hundred most dangerous professors in the United States—accurately warns us, "we need to fight for politics of direct democracy and direct action. The reinvigoration of the Left depends upon this" (p. 27).

Notes

An earlier version of this chapter appears in *Policy Futures in Education*. This is a profound, substantive, upgraded, and different version of the earlier piece in which I developed some of towering arguments and actually bring to the fore new challenges. I am in great debt to Sheila Macrine, David Hursh, Camille Martina, Donaldo Macedo, Jurjo Torres Santome, Mariano Enguita, Joe Kincheloe, Shirley Steinberg, Peter McLaren, Alvaro Hypolito, and Boaventura de Sousa Santos for some of the arguments raised here. Also I cannot forget my students at the University of Minho, Portugal, at the University of Coruna, Spain, and at Miami University, Ohio. To them my debt will be forever open. They alone know why.

1. For a detailed analyses of Michael Apple's work, see Paraskeva, 2004.
2. See Paraskeva, 2007a
3. PLC stands for public limited company. This concept is based on Ball's (2007) exhaustive and important study that tries to understand the effects of private sector participation in public sector education.

References

Agamben, G. (2005) *State of Exception*. Chicago, IL: University of Chicago Press.
Appadurai, A. (1996). *Modernity at Large: Cultural Dimensions of Globalization*. Minneapolis: University of Minnesota Press.
Apple, M. (2000). *Official Knowledge: Democratic Education in a Conservative Age*. New York: Routledge.
———. (2003). *The State and the Politics of Knowledge*. New York: Routledge.
Aronowitz, S. (2001). *The Last God Job in America*. Lanham. Rowman & Littlefield Publishers, Inc.
Arrighi, G. (1994). *The Long Twentieth Century: Money, Power, and the Origins of Our Times*. London: Verso.
Ball, S. (2001, Jul–Dez). Diretrizes Politicas Globais e Relações Políticas Locais em Educação. *Revista Curriculo Sem Ronteiras*, http:///29, 99–116, www.curriculosemfronteiras.org.

Ball, S.. (2005). *Educação à Venda*. Discursos: Cadernos de Politica Educativa e Curricular. Viseu: Livraria Pretexto Editora.
———. (2007). *Education PLC*. London: Routledge.
Bartolome, L. (2007). *Ideologies in Education*. New York: Peter Lang.
Bauman, Z. (2004). Wasted lives. Cambridge: Polity.
Berliner, D., and Biddle, B. (1995). *The Manufactured Crisis: Myths, Fraud and the Attack on the America Public Schools*. Reading: Addison Wesley.
Bourdieu, P. (1996). *On Television*. New York: The New Press.
Brown, Ph., and Lauder, H. (2006). Globalization, knowledge and the myth of the magnet economy. In H. Lauder, Ph. Brown, J-A. Dillabough, and A. Halsey (eds.). *Education, Globalization and Social Change* (pp. 317–340). Oxford: Oxford University Press.
Chomsky, N. (1989). *Necessary Illusions: Thought Control in Democratic Societies*. Cambridge: South End Press.
———. (2002a). The media: an institutional analysis. In P. Mitchell and J. Schoeffel (eds.). *Understanding Power: The Indispensable Chomsky* (pp. 12–15). New York: The New Press.
———. (2002b). Testing the propaganda model. In P. Mitchell and J. Schoeffel (eds.). *Understanding Power: The Indispensable Chomsky* (pp. 15–18). New York: The New Press.
Clarke, John E., and Newman, J. (1997). *The Managerial State*. London: Sage.
Clarke, John, Gewirtz, S., and McLaughlin, E. (2001). Reinventing the welfare state. In John Clarke, S. Gewirtz, and E. McLaughlin (eds.). *New Managerialism. New Welfare?* London: The Open University Press.
Cox, R. (2002). *The Political Economy of a Plural World. Critical Reflections on Power, Morals and Civilization*. London: Routledge.
Couto, M. (2005). *Pensatempos*. Lisboa: Caminho.
Derrida, J. (2002). Artifactuality, homohegemony. In J. Derrida and B. Stiegler (eds). *Echographies of Television* (pp. 41–55). Cambridge, England: Polity.
Eldridge, J. (1993). *Getting the Message: News, Truth and Power*. London: Routledge.
Ellison, R. (1952). *The Invisible Man*. New York: Modern Library.
Fairclough, N. (1992). *Discourse and Social Change*. Cambridge, England: Polity Press.
———. (2000). *New Labour, New Language*. London: Routledge.
Fiske, J. (1986). "Television: Polysemy and Popularity," Critical Studies in Mass Communication, *3*(4), 391–407.
Fiske, J., and Hartley, J. (1998). *Reading Television*. London: Methuen.
Freire, P. (1974). *Education: The Practice of Freedom*. London: Writers Readers Publishing Cooperative.
Ferguson, M. (2001). Why Radical Middle? Radical Middle Newsletter http://www.radicalmiddle.com/writers_n_pols.htm
Gabbard, D. (2003). Education *is* enforcement. the centrality of compulsory schooling in market societies. In K. Saltman and D. Gabbard (eds).

Education as Enforcement. The Militarization and Corporatization of Schools (pp. 61–78). New York: Routledge.

Gillborn, D. (1990). *"Race," Ethnicity, and Education: Teaching and Learning in Multi-Ethnic Schools.* London: RoutledgeFalmer.

———. (2004). Anti-racism: from policy to praxis. In Gloria Ladson-Billings and David Gillborn (eds.). *The RoutledgeFalmer Reader in Multicultural Education* (pp. 35–48). London: RoutledgeFalmer.

Giroux, H. (1981). *Ideology and Practice in Schooling.* Philadelphia: Temple University Press.

———. (1994). *Doing Cultural Studies: Youth and the Challenge of Pedagogy.* Retrieved August 2008 http://www.gseis.ucla.edu/courses/ed253a/giroux/giroux1.html.

Grossberg, L., Wartella, E., and Whitney, D. (1998). *Media Making. Mass Media in a Popular Culture.* London: Sage.

Hall, S. (1988). Marxism and the interpretations of culture. In C. Nelson and M. Grossberg (ed.). *The Toad in the Garden: Thatcherism among Theorists* (pp. 35–57). Urbana, IL: University of Illinois Press.

———. (1992). Cultural studies and its theoretical legacies. In L. Grossberg, C. Nelson, and P. Treichler (eds.). *Cultural Studies* (pp. 277–294). New York: Routledge.

———. (1996). "On postmodernism and articulation: an interview with Stuart Hall." In David Morley and Kuan-Hsing *Chen* (eds). *Stuart Hall: Critical Dialogues in Cultural Studies* (pp. 131–150). London: Routledge.

Herman S., and Chomsky, N. (2002). *Manufacturing Consent: The Political Economy of the Mass Media.* New York: Pantheon Books.

Hill. D. (2003). O neo-liberalismo global, a resistência e a deformação da educação. *Revista Currículo sem Fronteiras,* 3(2), 24–59. www.curriculosemfronteiras.org retrieved December 2006.

hooks, bell. (1994). *Outlaw Culture: Resisting Representations.* New York: Routledge.

House, E. (1998). *Schools for Sale, Why Free Markets Won't Improve America's Schools, And What Will.* New York: Teachers College Columbia University.

Hursh, D. (2006). Democracia sitiada. capitalismo global, neoliberalismo e educação. In J. Paraskeva, W. Ross, and D. Hursh (eds.). *Marxismo e Educação* (pp. 155–192). Porto: Profedições.

———. (2008). *High-Stakes Testing and the Decline of Teaching and Learning.* Lanham: Rowman and Littlefield Publishers, Inc.

Hutchison, R. (2006). *Their Kingdom Come: Inside the Secret World of Opus Dei.* NY, New York: St Martin's Press.

Jessop, B., Bonnett, K., Bromley, S., and Ling, Tom. (1984). Authoritarian populism, two nations and Thatcherism. *New Left Review,* 147, 33–60.

Kincheloe, J., and Steinberg, Sh. (2007). Cutting class in a dangerous era: a critical pedagogy of class awareness. In J. Kincheloe and Sh. Steinberg (eds.). *Cutting Class. Socio Economic Status and Education* (pp. 3–69). Lanham: Rowman and Littlefield.

Klein, Naomi. (2007). *The Shock Doctrine. The Rise of Disaster Capitalism.* New York: Metropolitan Books.

Kozol, J. (1999). *Amazing Grace: The Lives of Children and the Conscience of a Nation.* NY: Bt Bound.

Laclau, E., and Mouffe, C. (1985). *Hegemony and Socialist Strategy: Towards a Radical Democratic Politics.* London: Verso.

Lipman, P. (1998). *Race, Class and Power in School Restructuring.* New York: State University of New York Press.

Macedo, D., and Bartolomé, L. (2001). *Dancing with Bigtory: Beyond the Politics of Tolerance.* New York: Palgrave.

Macedo, D. (2006). *Literacies of Power: What Americans are Not Allowed to Know.* Boulder: Westview Press.

Macedo, D., Dendrinos, B., and Gounari, P. (2003). *The Hegemony of English.* Boulder: Paradigm Publishers.

Macrine, Sh. (2003). Imprision minds. The violence of neoliberal education or "I am not for sale." In K. Saltman and D. Gabbard (eds). *Education as Enforcement. The Militarization and Corporatization of Schools* (pp. 205–211). New York: Routledge.

McChesney, R. (1999). *Profit Over People: Neoliberalism and Global Order.* New York: Seven Stories Press.

———. (2004). *The Problem of the Media. U.S. Communication Politics in the 21st Century.* New York: Monthly Review Press.

McLaren, P., and Farahmandpur, R. (2002). Freire, Marx and the new imperialism: towards a revolutionary praxis. In J. Slater, S. Fain, and C. Rossatto (eds). *The Freirean Legacy—Education for Social Justice* (pp. 37–56). New York: Peter Lang.

Molnar, A. (1996a). *Giving Kids to Business. The Commercialization of America Schools.* Boulder: Westview.

———. (1996b). An interview with Alex Molnar, by Jay Huber. *Stay Free.* Retrieved October 17, 2006, http://www.stayfreemagazine.org/index.html.

Mouffe, C. (2000). *The Democratic Paradox.* London: Routledge.

———. (2006). *Democracia Agonística. Discuros—Cadernos de Politica Educativa e Curricular.* Viseu: Livararia Pretexto Editora.

Negri, A. (2002). *Império e Multitude na Guerra: O Imperio em Guerra.* Lisboa: Campo das Letras.

Newman, J. (2001). *Modernising Governance.* London: Sage.

Nussbaum, M. (1997). *Cultivating Humanity: A Classical Defense of Reform in Liberal Education.* Cambridge, Mass.: Harvard University Press.

Nyerere, J. (1998). Good governance for Africa. *Marxism and Anti-Imperialism in Africa*, www.marxists.ort. Retrieved December 2006.

Olssen, M. (2004). Neoliberalism, globalization, democracy: challenges for education. *Globalization, Societies and Education*, 2(2), 238–273.

Paraskeva, J. (2001). *As Dinâmicas dos Conflitos Ideológicos e Culturais na Fundamentação do Currículo.* Porto: ASA.

———. (2003). Desescolarização. Genotexto e Fenotexto das Políticas Educativas Neoliberais. In J. Torres Santomé, J. Paraskeva, and M. Apple (eds). *Ventos de Desescolarizacao: A Nova Ameaça à Escolarização* (pp. 91–107). Lisboa, Portugal: Editora Platano.

———. (2004). *Here I Stand: A Long Revolution. Ideology, Culture and Curriculum. Michael Apple and Critical Progressive Studies.* Braga: University of Minho, Portugal, E.U.

———. (2005). Portugal will always be an African Nation. A Calibanian prosperous or a prospering Caliban? In Donaldo Macedo and Panayota Gounari (eds). *The Globalization of Racism* (pp. 241–268). Boulder: Paradigm Press.

———. (2006). Neo-Marxismo com Garantias. In J. Paraskeva, W. Ross, and D. Hursh (eds). *Marxismo e Educacao. Volume I* (pp. 11–23). Porto: Profedicoes.

———. (2007a). Kidnapping public schooling: perversion and normalization of the discursive bases within the epicenter of New Right educational policies. *Policy Futures in Education*, 5(2), 137–159.

———. (2007b). Putting reality together. In D. Macedo and S. Steinberg (eds). *Media Literacy* (pp. 206–214). New York: Peter Lang.

———. (2008). *Mercantilizacao da Educacao.* Mangualde: Edicoes Pedago.

Said, E. (1997). *Covering Islam: How the Media and the Experts Determine How We See the Rest of the World.* New York: Vintage Books.

Saramago, J. (2003). O Nome e a Coisa. Julho: *Le Monde Diplomatique.*

Schmitt, C. ([1922] 2005). *Political Theology: Four Chapters on the Concept of Sovereignty.* Chicago, IL: University of Chicago Press.

Sen, A. (1999). Democracy as a universal value. *Journal of Democracy*, 10(3), 3–17.

Shivji, I. (2003). The struggle for democracy. *Marxism and Anti-Imperialism in Africa.* www.marxists.ort. Retrieved December 2006.

Sommers, M. (2000). Fear and loathing of the public sphere and the naturalization of civil society: how neoliberalism outwits the rest of us. *Paper Presented at the Haven Center Conferences Series*, University of Wisconsin, Madison, USA.

Sousa Santos, B. (1997). *Um Discurso sobre as Ciencias.* Oporto: Afrontamento.

———. (1998). *Reinventar a democracia.* Lisboa: Gradiva.

———. (2004). A critique of lazy reason: against the waste of experience. In I. Wallerstein (ed.). *The Modern World System in the Longue Duree* (pp. 157–197). Boulder: Paradigm Publishers.

———. (2005). *Another Knowledge is Possible.* London: Verso.

Sousa Santos, B., Nunes, J., and Meneses, M. (2007). Open up the cannon of knowledge and recognition of difference. In B. Sousa Santos (ed.). *Another Knowledge is Possible* (pp. ix–lxii). London: Verso.

Steinberg, S., and Kincheloe, J. (2001). Setting the context for critical multi/interculturalism: the power blocs of class elitism, white supremacy, and patriarchy. In S. Steinberg (ed.). *Multicultural Conversations. A Reader* (pp. 3–30). New York: Peter Lang.

Todorov, T. (2003). *A Nova Ordem Mundial.* Porto. ASA.

Torfing, J. (1999). *New Theories of Discourse.* Oxford: Blackwell.

Torres Santomé, J. (2001). Educacion en tiempos de neoliberalismo. Madrid: Morata.

Valenzuela, A. (1999). *Subtractive Schooling: U.S.-Mexican Youth and the Politics of Caring.* New York: State University of New York Press.

Vavi, Zwelinzima. (2004). Democracy has by-pass the poor. *Marxism and Anti-Imperialism in Africa.* www.marxists.ort. Retrieved December 2006.

Weiner, E. (2005). *Private Learning, Public Needs: The Neoliberal Assault on Democratic Education.* New York: Peter Lang.

Whitty, G. (2002). *Education and the Middle Class.* London: Open University Press.

Whitty, G., and Power, S. (2002, Jan.–Jun.). A escola, o Estado e o Mercado? A Investigacao Actualizada. *Revista Curriculo Sem Fronterias*, 2(1), 15–40. www.curriculosemfronteiras.org.

Williams, R. (1974). *Television. Technology and Cultural Form.* Ed. E. Williams. London: Routledge.

XV Governo Constitucional de Portugal. (2001). *Programa do XV Governo Constitucional Portugues.* Retrieved from www.portugal.gov.pt.

Chapter 8

Critical Teaching as the Counter-Hegemony to Neoliberalism

John Smyth

Introduction

Contemporarily in Western countries, teachers in public schools have been under siege. In policy terms, they have been treated in distrustful, demeaning, and exceedingly disrespectful ways by governments that through neoliberal policies overwhelmingly construct the work of teachers to serve the economy. Teachers have been told in no uncertain terms that their professional judgments are unwelcome, and that they are servants of capitalism at the behest of complicit and compliant governments. The relays through which this works are readily on display—high-stakes testing; distributing payment by performance or results; ranking and rating schools by means of league tables; publicly "naming and shaming" schools that are underachieving; penalizing schools through linking resourcing to notions of so-called school choice; distorting and trivializing the work of schools by requiring them to compete for "clients" through self-marketing and fake forms of image and impression management; technicizing the work of teachers through teacher-proof curricula and "scripted" forms of teaching; requiring schools to prostitute themselves to the corporate sector through pursuing various forms of sponsorship; and generally unleashing savage and unremitting "culture wars" against schools as public institutions and eulogizing the alleged benefits of all manner of privatisms. Taken collectively, these form an evil and devastating ensemble of tendencies that strike at the heart of that for which schools exist. They are grossly distorting and deforming, they are highly exploitative, demonstrably undemocratic, and they do not bode well for societies that purport to be civilized, enlightened, inclusive, and progressive—quite the contrary.

The argument of this chapter, consistent with the theme of this book, is that teachers can "speak back," and in many instances they are doing just that (see, e.g., Bigelow and Peterson, 2002). The starting point is the recognition that the self-interested agenda of capitalism, as given expression through neoliberal policy trajectories and discourses, is an "uncouth interloper . . . arrogant, alien and improper" (Adorno, 1974, p. 23). Spaces exist within the work of teaching in which teachers can exercise agency, through the way they can work with young people to unveil and unmask how power works and puncture the mythology that individualism, competition, and consumerism are the only, or indeed the best, alternative that is available. Such approaches are not "quick fix" solutions in the way that neoliberalism presents itself, but rather they are long-term ideals or aspirations to be worked toward to create a more tolerant, equitable, and just society.

I will not labor the point any longer here about the forces operating to construct teachers as technicians and docile servants except to say that those forces are pervasive, persistent, and unlikely to go away of their own accord any time soon. The more optimistic, hopeful, and constructive possibilities lie in scoping out a different imaginary for teachers, and within it, a practical perspective by which they might advance in that direction. Myself and others have labeled this "critical teaching" (Kincheloe and Steinberg, 1998; Shor, 1980, 1992; Shor and Pari, 1999; Smyth, 2000), although other terms capture it equally well, like critical pedagogy (Duncan-Andrade and Morrell, 2008; Kincheloe, 2004; Wink, 1997), teaching for social justice (Ayers, 2004; Shor, 1987), and critical education (Cooper and White, 2006; Hinchey, 1997, 2004). In the preface to his book *Education is Politics* (Shor and Pari, 1999), Shor captured the essence of the current struggle through a metaphor:

> Two great rivers of reform are flowing in opposite directions across the immense landscapes of American education. One river flows from the top down and the other from the bottom up. The top-down river has been the voice of authority proposing conservative agendas that support inequality and traditional teaching; the bottom-up flow contains multicultural voices speaking for social justice and alternative methods. (P. vii)

Shor also points out that within this rather confusing situation there is a profound paradox, in which at one and the same time schools are: "becoming more authoritarian and more unequal, more unstable and damaging for children, thus provoking action from below by teachers, parents, community coalitions, organized advocates, and national networks" (p. ix). With the benefit of hindsight, Shor might have been somewhat overly optimistic in his assessment that "popular outrage at and mass alienation from the system is perhaps also at its peak," but he may be closer to the mark with his comment that "a long-term showdown is in the making" (p. ix). Hopefully, this chapter is a modest contribution to that turnaround.

Teachers as Intellectuals/Political Actors

The notion that, perhaps, teachers are more than mindless technical operatives has been expressed widely and forcefully going back to Dewey and earlier. But, as Warham (1993) noted, no matter how vigorously one tries to escape from it, teaching is by nature a political activity, both inside the classroom and outside of it, by virtue of those forces trying to make it otherwise. She argues that in order for a theory of teaching to be viable it must acknowledge this undeniable fact. Probably more than any other contemporary educator, Giroux (1988) has championed the cause of teachers as intellectuals and the political meaning behind that role. Giroux's extensive writings (1985a) have occurred against a backdrop of educational reforms that have aimed to reduce teachers to "the status of low-level employees or civil servants whose main function seems to be to implement reforms decided by experts in the upper levels of state and educational bureaucracies" (p. 20). The consequence, Giroux argues, is that the search and the push for technical/administrative solutions to the complex economy/society/education linkage has produced a growing gulf between those who decide on technical and methodological grounds what is best for schools, and the schools and teachers who deal with students, curricula, and pedagogy on a daily basis. He notes that there is a process of subjugation of intellectual labor at work here that in many cases reduces teachers to the "status of high level clerks implementing the orders of others...or to the status of specialised technicians" (p. 21).

This dominance of technocratic rationality has produced a form of proletarianization of teachers' work, not dissimilar to what happened to factory workers in the nineteenth century, as the control of what had previously been highly independent craftsman came increasingly under corporate and factory control. But, as Giroux (1985a) argues, what has occurred in schooling is more than just an elevation of the importance of the technical and the economic in the everyday life of schools:

> Underlying this technical rationality and its accompanying rationalisation of reason and nature [has been] a call for the separation of conception from execution, the standardisation of knowledge in the interests

> of managing and controlling it, and the devaluation of critical intellectual work for the primacy of practical considerations. (P. 23)

In a similar vein, teacher education has, Giroux says, all too often been reduced to questions of "what works" (see Bennett, 1986; and Berliner and Biddle, 1995, for a critique). Issues to do with what counts as knowledge, what is worth teaching, and how one judges the purpose and nature of teaching have thus become submerged (or even obliterated) in the press for routinization and standardization through what Giroux terms "management pedagogies."

Giroux's claim (1985a) is that one way of rethinking and restructuring teachers' work is to view teachers as intellectuals and to view what teachers do as a form of intellectual labor. The argument is that if teachers are seen in this light one can begin to "illuminate and recover the rather general notion that all human activity involves some form of thinking [and] no activity, regardless of how routinised it might become, is abstracted from the functioning of the mind in some capacity" (p. 27). Applying this notion to teaching, Kohl (1983) stated:

> I believe a teacher must be an intellectual as well as a practitioner... I don't mean an intellectual in the sense of being a university professor or having a PhD... I am talking about activities of the mind. We must think about children, and create many philosophies of life in the classroom... An intellectual is someone who knows about his or her field, has a wide breadth of knowledge about other aspects of the world, who uses experience to develop theory and questions theory on the basis of further experience. An intellectual is also someone who has the courage to question authority and who refuses to act counter to his/her experience and judgment. (P. 30)

Equally important is Kohl's argument that unless teachers themselves make theories and test them, theories will be made for them by researchers and other groups only too willing to move in and fill "the vacuum that teachers have created by... giving up their responsibility as intellectuals" (p. 30).

The kinds of questions that Giroux (1985b) suggests are necessary if teachers are to interrogate their work so as to become intellectuals include:

> What counts as school knowledge?
> How is such knowledge selected and organized?
> What are the underlying interests that structure the form and content of school knowledge?

> How is what counts as school knowledge transmitted?
> How is access to such knowledge determined?
> What cultural values and formations are legitimated by dominant forms of school knowledge?
> What cultural formations are disorganized and delegitimated by dominant forms of school knowledge? (P. 91)

Teachers need to be political actors in their educational settings by being clear about the different ways in which they experience their work—how they encounter it, how they understand it, and how they feel about it (Ginsburg, 1988, p. 363). Adopting a political stance to one's work does not mean being a political partisan; it involves what Popkewitz (1987) describes as "critical intellectual work." Here, "critical" means moving outside the assumptions and practices of the existing order, and struggling to make categories, assumptions, and practices of everyday life problematic (p. 350). But, as Ginsburg (1988) argues, it is more than just problematizing the work of teaching because it involves "struggle to challenge and transform the structural and cultural features we... come to understand as oppressive and anti-democratic" (pp. 363–365). Ginsburg's point is that teachers need to see themselves as actively participating in progressive movements committed to bringing about fundamental social change (Anyon, 2005). According to this view, the image of teachers as compliant, passive, and easily molded workers is replaced by a view of the teacher "as an active agent, constructing perspectives and choosing actions" (Feiman-Nemser and Floden, 1986, p. 523).

The construal of teaching as a form of intellectual labor would have one regard teachers' own classrooms and schools as sites of serious inquiry, with questions being asked and answered as to what schooling is about, how it works for some students, and what conditions act to exclude others. This view of teaching is political in the sense that teachers do not take the nature of their work for granted (see Carlson and Apple, 1998); they are prepared to question how it came to be that way and what sustains and maintains this set of views.

This has several dramatic and direct implications for students. The most obvious is a bringing of student lives, perspectives, cultures, and experiences into the center of curriculum in a way that involves students as coconstructors and cocreators (rather than passive consumers) of that curriculum, along with teachers. This preparedness of teachers and schools to take students' lives seriously means that the hierarchically scripted curriculum has to be modified (if not totally jettisoned) in order to accommodate the storied and narrative representations of the way students lead their increasingly complex lives (see, e.g., Ayers, Hunt, and Quinn, 1998). This undermining means that issues previously off limits will have to be brought to center stage and confronted in the classroom, issues such as racism, homophobia, gender, violence, poverty, and economic exploitation. It will also mean that students will be more prominent, through, for example, researching the contexts and communities in which they live, unmasking questions that are usually marginalized or pushed off the social and educational agenda of schooling. Thus, as teachers embrace the political in their work, a fundamental shift will occur in the direction of genuine sharing of power with students in ways that go considerably beyond many current inauthentic and tokenistic attempts.

Problematizing Approaches to Teaching

To problematize teaching involves challenging habits and methods taken for granted and questioning fundamental and cherished assumptions. This usually occurs through a collective and collaborative process of teachers working with one another, examining and interrogating their teaching with probing questions such as:

> What is happening here?
> Who says this is the way things ought to happen?
> Who is it that is defining the work of teaching?
> How is that definition being fought over and resisted in various ways?
> What concessions and accommodations are being made?
> How are issues of skill, competency, professionalism and autonomy being expressed in the social relations of teaching?
> Whose interests are being served in the change process?
> What new forms of power are being used to focus power relations in teaching?
> How are the re-defined labor relations of teaching being played out? (Smyth, 1995a, p. 85)

And:

> What overall purposes are being served?
> Whose vision is it anyway?
> Whose interests are being served?
> Whose needs are being met?
> Whose voices are being excluded, silenced, denied?
> How come some viewpoints always get heard?

Why is this particular initiative occurring now?
What alternatives have or should have been considered?
What kind of feasible and prudent action can we adopt?
Who can we enlist to support us?
How can we start now?
How are we going to know when we make a difference? (Smyth, 1995b)

Questioning of this kind moves significantly beyond examining teaching in limited individualistic frames or exclusively in terms of personal deficits (either of teachers or children). The starting point might be instances of classroom activity that are perplexing, confusing, or troublesome, and the intention is to locate or situate these in relation to wider change forces.

As long as attempts to improve teaching continue to be couched solely in terms of perceived individual deficits within teacher's pedagogical repertoires and styles (or the learning styles of students), the process of improving teaching will fail to grapple systematically with historical and structural factors that make teaching (and learning) the way it is. Constructing problems as individual deficits also obscures the class, race, and gender blindness of curricula and hence wider understandings of how power is exercised and in whose interests.

To problematize teaching is to adopt a "critical" approach, by which I mean the posture of "a certain scepticism, or suspension of assent, towards [an]...established norm or mode of doing things" (McPeck, 1981, p. 6). According to Garrison (1991), acting critically involves not taking things for granted, but it must also involve a constructive, positive, and proactive aspect of allowing for alternative possibilities through "a search for a more satisfactory insight or resolution of a troubling situation" (p. 289). Cox (1980) defined critical "in the sense that it stands apart from the prevailing order of the world and asks how that order came to be" (p. 230).

Regarding teaching from the position of seeing it problematically involves raising doubts about the existing state of affairs. In the context of the classroom, a critical stance means raising questions about how the wider social context shapes and informs what goes on through questions such as:

- Who talks in this classroom?
- Who gets the teacher's time?
- How is ability identified and attended to here, and what's the rationale?

- How is the unequal starting points of students dealt with here?
- How are instances of disruptive behavior explained and handled?
- Is there a competitive or a co-operative ethos in this classroom?
- Who helps who, here?
- Whose ideas are the most important or count most?
- How do we know that learning is occurring, here?
- Are answers or question more important in this classroom?
- How are decisions made here?
- How does the arrangement of the room help or hinder learning?
- Who benefits and who is disadvantaged in this classroom?
- How is conflict resolved?
- How are rules determined?
- How are inequalities recognized and dealt with?
- Where do learning materials come from?
- By what means are resources distributed?
- What career aspirations do students have, and how is that manifested?
- Who determined standards, and how are they arrived at?
- How is failure defined, and to what is it ascribed? Who or what fails?
- Whose language prevails in the classroom?
- How does the teacher monitor his/her agenda?
- How does the teacher work to change oppressive structures in the classroom?
- What is it that is being measured and assessed in this classroom?
- Who do teachers choose to work with collaboratively, on what, and under what circumstances? (Smyth, 1995b)

Developing ways of "extraordinarily re-experiencing the ordinary" (Shor, 1980, p. 37) involves posing questions such as "why are we doing this?" and "where will it get us?" This might involve disturbing and uncovering the way domination acts in our daily lives and becoming increasingly vigilant about the contradictions we live.

For example, Connell (1994) has shown that "Australian schools deliver massive advantages to the children of the well-off and well-educated parents, and massive disadvantages to children of the poor and the poorly educated" (p. 1). That schools produce these effects over long time periods show that such circumstances "are not accidental, and they cannot be eliminated by minor tinkering with the system. It is clear that our education system is designed in a way that delivers 'success' along lines of social and economic privilege. The way school knowledge is defined, and school learning is organized, has this effect" (p. 1). The consequence is that where education produces success for the privileged, education overall as a social institution fails—because the moral nature of the enterprise converts teachers into "gatekeepers of privilege" implementing curriculum that "narrows the immense possibilities of learning" (ibid.). This tendency is exacerbated by the already well-advanced moves toward privatizing public education brought on by the exhortation to teachers, students, and schools to compete against one another. This is a recipe for social stratification, segmentation, and alienation.

Connell (1998) points to the well-acknowledged taken-for-granted "Competitive Academic Curriculum" (CAC), with its "abstract division of knowledge into 'subjects'; a hierarchy of subjects; a hierarchical ordering of knowledge within each subject; a teacher-centered classroom-based pedagogy; an individualized learning process; [and] formal competitive assessment (the 'exam')" (p. 84). This curriculum is constructed and implemented in ways that serve the interests of the *most* advantaged students while failing to value the experiences, speech, ideas, and skills of students from the most disadvantaged groups in society. In other words, instead of constituting some set of common learnings for all students, the CAC acts as a device for selecting those students whose largely middle class backgrounds and experiences coincide with what is considered valuable and worthwhile knowledge. For the remaining students, mostly from minority and disadvantaged groups, whose life experiences and cultures differ dramatically from the dominant (mostly male, Anglo, middle class), the effect is subordination, marginalization, and eventually failure (Connell, 1993).

Connell (1994) argues the need to move beyond compensatory approaches for targeted groups, working instead for positive social justice at the grassroots level, through posing questions and pursuing action around:

- What way of organizing learning and teaching will most benefit the least advantaged?
- What concept of teacher professionalism will most benefit the least advantaged?
- How can we democratize the relationship between public schools and working-class communities?
- How can we redefine the relationships between schools and other public and cultural institutions? (p. 2)

Questions like these require confronting the oppressive nature of individualism in society and seeing that students actually come from social groups. At the moment, however, schools and classrooms are deeply entrenched in a pervasive ideology that frames modes of thinking and acting exclusively in terms of measurement and assessment, school discipline and behavior management, competencies, standards, and the like. Moving outside of this frame requires teachers who are able to uncover the dominant fallacy that "individuals are autonomous and [therefore] responsible for making their own way," and that failure is something that can therefore be "attributed to personal not social causation" (Razack, 1993, p. 49). The individual rights model does not offer any conceptual tools for understanding oppression and the systems that constrain individual choice. The model for positive social justice therefore shifts the focus to the systemic processes that block opportunities for groups, by exposing the inadequacy of individual rights models for explaining what occurs in classrooms.

Teachers therefore need to work in ways that challenge the taken-for-granted in their teaching and operate from the position that there may be other more just, inclusive, and democratic ways that overcome various forms of classroom disadvantage. This means a preparedness by teachers to challenge the status quo (Smyth, Hattam, and Lawson, 1998).

Teaching for Social Responsibility and against the Grain

Friedberg (1992) helpfully highlights the central issue in classrooms as being how to convert the "power of position" into the "expertise of authority." He says doing this requires questioning:

> What would an education based on dialogue and empowerment look like across cultures with their various ways of engaging in talk and learning? What would it mean to take democracy seriously in our daily lives? What kinds of relationships must exist among teachers so that they can respond directly and forcefully to racism, sexism and other forms of fascism, and the violence they bring, without simultaneously disempowering others? What are the possibilities for dialogue in a society that seems to value "talking at" more than "talking with"? (P. 212)

For Friedberg, the major obstacle is the "coercive character of schooling, and in particular the traditional power of the teacher" (p. 154). Relationships of domination saturate what occurs in schools, and the greatest challenge lies in developing in teachers the capacity to enter into "power with relationships" with their students. Achieving this may be extraordinarily difficult because the lived relationships of teachers, the way they think about themselves and how they connect to other teachers, may itself constitute a major form of domination.

Warham (1993) cast some light on what she regards as teachers employing dominant strategies, such as when they use: ritual language; routines to control large groups; loud tones of voice; nonverbal communication to control or assert power; and the manipulation of the structure of classroom discourses to exclude certain students (p. 214). On the other hand, teachers use less dominant strategies when they: utilize peer group pressure, encourage students, establish long eye contact, manipulate classroom discourse to support and include children, encourage positive thinking, ask favors, create group coherence, restate what children say when they lose track of the discourse, keep quiet and allow children to take responsibility for a discussion, engage in acts of politeness and kindness, make suggestions, praise and provide compliments, and enjoy humor (p. 214).

Classroom power, however, is never uni-directional; it is always two-way and dialectical, with resistance (in varying degrees) from and by students. When students use dominant strategies on their teachers they: do not pay attention; yawn and show disinterest; demand attention; do not listen to their teacher; fiddle and distract other children; shuffle and become restless; and disobey the teacher's instructions (p. 215). When students use less dominant power strategies on the teacher, they establish long eye contact, smile, enjoy humor, cooperate, listen carefully, and obey instructions (ibid.). Within and between the kind of spaces alluded to, it becomes clear that learning is not what the teacher/school/system requires that students do, so much as a negotiated process "of what the children will allow the teacher to do" (ibid.).

Eichler (1992) argues that in terms of barriers to teachers operating critically, there is much in the educational literature that amounts to "bogus teacher development" (p. 3) that prevents teachers from asking "questions related to the broader purpose of education in a democratic society" (ibid.). Continuing to perpetuate "isolation of individual teachers and the lack of attention to the social context of teaching" (p. 9) means that this "individualist bias makes it less likely that teachers will be able to confront and transform those structural aspects of their work that hinder the accomplishment of their educational mission" (p. 8). Eichler is illustrative of someone who seeks to engage teachers with questions about how their "everyday actions challenge or support various oppressions and injustices related to social class, race, gender, sexual preference, religion and numerous other factors..." (p. 12). His point is that we cannot assume teachers' willingness to educate everyone's children, and unless teachers, even innovative ones, develop a commitment "to the quality of relationships" and work to ensure that "everybody's knowledge and cultural heritage is represented," then "many students will continue to be bypassed by innovative school practices and continue to be denied, with the complicity of the school, access to decent and fulfilling lives" (ibid.). While it is unacceptable, he says, that teaching be reduced to "only its political elements" (p. 14), it is equally unacceptable that issues of equity and social justice be relegated to the category of other people's problems outside of the classroom.

Cochran-Smith (1991) and Simon (1992) have also pursued the idea that to bring about change through teaching, it is necessary that teachers place themselves in situations in which they work against and challenge established norms. For Simon, the essence of changing teaching lies in trying not to succumb to pressures to rigidly encase notions such as "critical pedagogy" within definitions or procedures, but rather to see them as being reference points for "an ongoing project and certainly not a prescriptive set of practices" (p. xvi). Simon presumes that "teachers are cultural [and political] workers" and, as such, they engage in a process of helping students, "challenge[in] and assess[in] existing social conventions, modes of thought, and relations of power" (p. 35) and arrive at conclusions about how situations came to be the way they are. He therefore admits to having a political vision that structures his work, that embraces questions such as:

> How is experience to be understood? What information and experience do I have access to that is important and possibly helpful to others? In what way does the form and substance of the knowledge engaged in teaching situations enable/constrain personal and social possibilities? How do I understand learning and the relationship teaching has to it? What is my view of a "person"? How do emotions, desires, and psychic investments influence teaching situations? How do the oppressive forms of power in my community manifest themselves in classrooms, and how do I situate myself in relation to such forms? And finally, how do I define my responsibilities as a teacher—to what should my students be held accountable, and to what should I be held accountable? (P. xvi)

Simon is seeking to offer "a critical yet constructive way of reconstituting educational practice" (p. 5). All the while he keeps in mind, in the best tradition of self-reflection, that his own experiences as a teacher are continually woven into his "participation in existing relations of domination" (p. 6). By focusing on his own contradictions he can come to "know where there is work yet to be done" (ibid.).

Cochran-Smith (1991) claims that what distinguishes schools and teachers that work "against the grain" from others is their strong commitment to interrupting conservative influences and developing structures that produce "critical dissonance" and "collaborative resonance" (p. 304). That is to say, they generate "intensification of opportunities to learn from teaching through the co-labor of communities" (p. 304) that focus on how school participants themselves might begin to "bear upon the institutional and instructional arrangements of schooling" (p. 282). The process of collaborative resonance involves teachers in "critiqu[ing] the cultures of teaching and schooling, research[ing] their own practices, articulat[ing] their own expertise, and call[ing] into question the policies and language of schooling that are taken for granted" (p. 283).

Teaching for social responsibility, as it comes through in the writing of people like Kriesberg, Zeichner, Simon and Cochran-Smith, underscores aspects such as:

- Converting positional power into expertise of authority.
- Teachers working through dialogue with students.
- Teachers fostering forms of teaching that transcend dominance.
- Asking questions about the broader purpose of teaching.
- Moving outside of isolated teaching arrangements.
- Engaging with questions about how everyday teaching supports injustices and oppression.
- Assisting students to challenge existing social conventions and arrangements.
- Above all, having a moral vision for schooling.

Teaching for Democracy and Social Justice

Engaging students with the big questions that fire the imagination and the forces that shape their lives is what Wood (1990) regards as the hallmark of a socially critical teacher. For example, in areas of high unemployment, teachers "as curriculum workers" might engage their students with questions such as: "what work is here?; why are there no industries?; how can we get higher unemployment benefits?" While questions such as these do not have ready answers, they are a starting point for students to see beyond victim-blaming responses to what is occurring. In these ways, Wood's agenda is to have teachers develop critical forms of literacy that involve students in "the ability to evaluate what is read or heard with respect to the interest being served" (Wood, 1988, p. 178). In particular, "critical literacy involves building reading skills around students' own reading agendas. Having them read about things in which they have an interest and helping them write their own reading material are key components in this process" (p. 178). Viewed in this way, the curriculum Wood (1990) argues, becomes "shared—not...something that teachers dispense the way physicians prescribe medicine. Curriculum is a process in which teachers and students engage to order and make sense of the world" (p. 107).

Through the practices of Harmony School, Goodman (1992) highlights the tension between "individuality" and "community" as the school sought to rework what was meant by the notion of critical democracy. The experiences of Harmony show the importance of enabling students to acquire a connectionist perspective toward power that is tied to "increasing their sense of social responsibility." Goodman argues that the reciprocity that comes with power is crucial to its democratic enactment. The struggle is between a societal ethos of competitive individualism and a school that is trying to develop a collective sense of caring and a commitment to shared responsibility. Goodman's "connectionist" view of power rests upon the claim that students must be given genuine opportunities to negotiate successfully with their teachers, and that in the process, teachers must be put "at the center of the curriculum." This is a far cry from worldwide moves at the moment to re-centralize control over schools, to colonize the curriculum, to narrow the curriculum through uniformity, and to turn schools into institutions preoccupied with cultural production and mindless vocational training of the worst kind.

The central organizing image Goodman describes is the importance of "community" in schooling. It is the "dialectical tension" between an atomistic society on the one hand, and one characterized by social conformity on the other, and the tendency for these to get out of balance that is crucial. Goodman's highlighting of how individualism reaches deep into American culture makes it easier to comprehend the significance of the attempt by teachers, administrators, parents, and students at Harmony to work "against the grain."

One of the recurring themes in Goodman's book is that while teachers at Harmony mostly had a well-developed sense of caring and social responsibility in the way they worked with their students, they had been untouched by the forms of critical theorizing emanating from the academy. This is both understandable and explainable given the genre and the terrain upon which the largely academic debate has occurred. The difficulty for schools like Harmony is in the lack of connectedness to wider arenas of struggle outside of schools (see Anyon, 2005). This may be in part due to the need for those of us outside of schools to find better ways of "adopting" practitioners in schools and working with them to better theorize the essence of what is occurring in schools. Goodman's description of what was being attempted at Harmony school, rather than providing a recipe to be followed, acts more as a beacon to work toward. Notions of democratic schooling are not forms of dogma to be blindly followed or transferred from one setting to another—rather, they are social, cultural, and political constructions that reflect the way in which people inside schools choose to live their lives as students, teachers, administrators, and parents.

Critical teaching for democracy and social justice can take many forms and shapes, and we need to not lose sight of what these approaches mean practically speaking for students. Here is and extended example of what this might look like.

Bigelow (1998), a U.S. classroom teacher, has his students look "Behind the labels, the global sweatshop, Nike, and the race to the bottom":

> I began the lesson with a beat-up soccer ball. The ball sat balanced on a plastic container in the middle of the circle of student desks. "I'd like you to write a description of this soccer ball," I told my high school Global Studies class. "Feel free to get up and look at it. There is no right or wrong. Just describe the ball however you'd like."
>
> Looks of puzzlement and annoyance greeted me. "It's just a soccer ball," someone said... "I'm not asking for an essay," I said, "just a paragraph or two."
>
> As I'd anticipated, their accounts were straightforward—accurate if uninspired. Few students accepted the offer to examine the ball up close. A soccer ball is a soccer ball. The sat and wrote. Afterwards, a few students read their descriptions aloud. Brian's is typical: "The ball is a sphere which has white hexagons and black pentagons. The black pentagons contain red stars, sloppily outlined in silver...One of the hexagons contains a green rabbit wearing a soccer uniform with 'Euro 88' written parallel to the rabbit's body. This hexagon seems to be cracking. Another hexagon has the number 32 in green standing for the number of patches that the ball contains."
>
> But something was missing. There was a deeper social reality associated with this ball—a reality that advertising and the consumption-oriented rhythms of U.S. daily life discouraged the students from considering. "Made in Pakistan" was stenciled in small print on the ball, but very few students thought that significant enough to include in their descriptions. However, these three tiny words offered the most important clue to the human lives hidden in "just a soccer ball"—a clue to the invisible Pakistanis whose hands crafted the ball sitting in the middle of the room. (Pp. 21–22)

After some discussion about their writing and after reading a poem to reorient them, Bigelow invited his students to "resee" the soccer ball:

> If you like, you can write from the point of view of the ball, you can ask the ball questions, but I want you to look at it deeply. What did we miss the first time around? It's not "just a soccer ball." With not much more than these words for guidance—although students had some familiarity with working conditions in poor countries—they drew the line beneath their original descriptions and began again.
>
> Versions one and two were like night and day... Pakistan as the country of origin became more important. Tim wrote in part: "Who built this soccer ball? The ball answers with Pakistan. There are no real names, just labels. Where did the real people go after it was made?" Nicole also posed questions: "If this ball could talk, what kinds of things would it be able to tell you? It would tell me about the lives of the people who made it in Pakistan... But if it could talk, would you listen?"... And Sarah imagined herself as the soccer ball worker. "I sew together these shapes of leather. I stab my finger with my needle. I feel a small pain, but nothing much because my fingers are so callused. Every day I sew these soccer balls together for 5 cents, but I've never once had a chance to play soccer with my friends. I sew and sew all day long to have these balls shipped to another place where they represent fun. Here, they represent the hard work of everyday life." When students began to consider the human lives behind the ball-as-object, their writing also came alive... Students had begun to imagine the humanity inside the ball; their pieces were vivid and curious. The importance of making the visible the invisible, of looking behind the masks presented by everyday consumer goods, became the central theme of my first-time effort to teach about the "global sweatshop" and child labor in poor countries. (Pp. 22–23)

Needless to say, Bigelow was unsuccessful in getting the nearby representative from Nike to come and dialogue with his students. But he did have his students engage in a "global clothes hunt"—bringing items of clothing or toys from hone (T shirts, pants, skirts, shoes, Barbie dolls, baseball bats, etc.) to do an analysis of where the items were made and what sense the students made of this. He displayed their maps and collages and had the class search for patterns regarding where the goods were produced:

> Some students noticed that electronic toys tended to be produced in Taiwan or Korea; that more expensive shoes, like Doc Martens, were manufactured in Great Britain or Italy; athletic shoes were made mostly in Indonesia or China. On their "finding patterns" write up,

> just about everyone commented that China was the country that appeared most frequently on people's lists. A few kids noted that most of the people in the manufacturing countries were not white. As Sandee wrote, "The more expensive products seemed to be manufactured in countries with a higher number of white people. Cheaper products are often from places with other than white." We'd spent the early part of the year studying European colonialism, and some students noticed that many of the manufacturing countries were former colonies. I wanted student to see that every time they put clothes on or kick a soccer ball they are making a connection, if hidden, with people around the world—especially in the Third World—and that these countries are rooted in historical patterns of global inequality. (P. 28)

Bigelow's examples beautifully capture the essence of what is involved in critical teaching emerging from everyday life. This perspective can be summarized as follows:

- teachers engaging students with questions that have relevance beyond the classroom;
- working with students in ways that enable them to delve more deeply into content that is normally presented to them;
- schools and teachers operating in other than individual and competitive ways, and creating forms of shared responsibility and community;
- changing mindsets and orientations rather than using "how-to-do-it" approaches;
- listening to voices that originate from within classrooms;
- Using personal experience as a starting point and source of knowledge;
- questioning the authority of the teacher as the sole source of knowledge;
- students themselves becoming important alternative sources of theorizing about learning;
- focusing on how power is reproduced through structures and forms of language;
- encouraging the translation of democratic processes pursued inside the classroom into venues outside.

Critical Teaching and Critical Pedagogy in the Classroom

Shor's *Critical Teaching and Everyday Life* (1987) and *When Students Have Power* (1996) explore ways of democratizing social relations in the classroom that are indicative of what it means to operate in dialogical ways. He shows how sources within our culture interfere with critical thought. His starting point is schooling as a social practice where "dialogue is a democratic model of social relations, used to probematize the undemocratic quality of social life" (1987, p. 95).

Bigelow (1990, 1992) pursues a similar agenda in creating a classroom that becomes "part of a protracted argument for the viability of a critical and participatory democracy" (1992, p. 19). Bigelow seeks to get his students to see how society reproduces inequalities. He does this through a "dialogical" approach to teaching in which he has his students critique the larger society by probing the social factors that make and limit their lives, who they are, and who they *could* be (pp. 19–20). Bigelow's concern is to confront the dynamics of power and the role of resistance—in other words, learning about the "causes for their own insubordination, [and] the role they could play…in resisting it" (p. 22). Bigelow (1992) is clearly "excited by this sociological detective work" (p. 22) in which he sees his own role as helping shape students' perceptions of the larger society.

Viewed in this way, schooling becomes a project of helping students to see injustices and assisting them both to locate themselves in relation to such issues and to see how society is structured in ways that both sustain and maintain those inequities. This necessarily involves working with students in ways that enable teachers to see through their own pedagogical work in classrooms, how knowledge and power are inextricably linked, and that students also need to see themselves as having a crucial role in "problematizing knowledge, utilizing dialogue, and making knowledge meaningful so as to make it critical in order to make it emancipator" (Giroux, 1985a, p. 87).

My own approaches (Smyth, 1991) in *Teachers as Collaborative Learners: Challenging Dominant Forms of Supervision* seek to engage teachers in recognizing and extirpating inappropriate pedagogies and ideologies. Instead of the toxic practices being pressed upon schools from the corporate sector that are producing such distortion and disfigurement, schools need coherent theories about teaching crafted by teachers themselves. I have envisaged a number of layered moments (Smyth, 1993):

Describing

If teachers are to celebrate the virtues of their work, as well as convince others of its efficacy, then the starting point is to describe the situational specifics of their teaching. In this teachers need the assistance of colleagues in looking for similarities, differences, patterns, regularities, discontinuities, contradictions, and ruptures in their teaching. In a word, they need assistance in becoming "informed" so as to see the theories in their teaching.

Informing

Other people's theories are often foisted onto teachers. But, given the opportunity of explaining the meaning that lies behind their teaching, teachers can be perceptive at unraveling the complexity of classrooms. They are thus able to unravel their teaching so as to exemplify local theories and gain a platform from which to make sense of their teaching as well as from which to explain it to others. No longer having to depend on others' theories also provides teachers with the courage to confront whoever is doing the defining, articulating, and legitimating knowledge about teaching. The issue of who has the right to do these things can be confronted and therefore contested.

Confronting

Teachers can ascertain "how things came to be this way" and what broader forces operate to make them like this biographically by pursuing questions such as:

- What do my teaching practices say about my assumptions, values, and beliefs about teaching?
- What social practices are expressed in these ideas?
- What causes me to maintain my theories?
- What constrains my views of what is possible in teaching?

By the time teachers have begun to grapple with questions like this and the forces that are shaping what they do, they are starting to think about how to act in different ways and are moving toward *reconstructing* parts of their teaching and the contexts and structures within which they do it.

The key thematic issues I have focused on through critical approaches to teaching and classroom pedagogy have included:

- asking critical questions as the major method by which teachers probe their teaching;
- challenging passivity;

- searching for interrelationships between narrated storied accounts of teaching, and the deeper meanings in teaching;
- critiquing the everyday practice of teaching, as well as envisaging what alternatives might look like, is crucial;
- asking how teaching becomes unwittingly implicated in maintaining the status quo;
- pursuing a number of moments or phases embracing elements of describing, informing, confronting, and reconstructing teaching.

Conclusion

The central point of this chapter has been that teaching for social responsibility, or acting in socially critical ways, is crucial in creating counter-hegemonic resistance to neoliberalism. Schools are one of the few remaining social institutions that still have a capacity to enculturate the young in ways of organization that celebrate social relationships. Teachers are crucial agents in the process of asking questions that unmask the nature of teaching and learning and the interests being served or denied.

I have explored an issue that rarely gets coverage in debates about teaching and teachers' work—namely, teachers as social theorists and political actors. My broader purpose has been to illustrate how teaching might be construed so as to assist in the reclamation of a democracy of social responsibility, and engaging teachers in investigating the practice, meaning, and intent of what lies behind schooling. The central argument has been that for too long teachers have been treated in ways that deny the importance of teaching as a social practice—one concerned with teaching for social responsibility, democracy, social justice, connectedness, and civility. Teachers' work has always been an avowedly political process, long characterized by decisions about: what knowledge gets taught, and what gets omitted; whose view of the world is privileged, and whose is denied; what forms of pedagogy are inclusive, and which are exclusive; and, whose interests are served, and whose are marginalized and excluded. These are no longer matters that should be spoken about in hushed tones, for it is clear that if teachers are not political about their work (in the sense of being critically reflective about it, and the implications that has for the life chances of children), then they are the only group affiliated with teaching who operate in such allegedly detached ways.

Note

Publication of this paper was made possible through research originally funded by a grant from the Australian Research Council. Some of the ideas and arguments here are a condensation, abridged, and significantly re-worked version of my "Reclaiming social capital through critical teaching," *The Elementary School Journal*, 100(5), 2000, 491–511.

References

Adorno, T. (1974). *Minima Moralia: Reflections from Damaged Life.* London: Verso.

Anyon, J. (2005). *Radical Possibilities: Public Policy, Urban Education and a New Social Movement.* New York: Routledge.

Ayers, W. (2004). *Teaching the Personal and the Political: Essays on Hope and Justice.* New York: Teachers College Press.

Ayers, W., Hunt, J., and Quinn, T. (1998). *Teaching for Social Justice: A Democracy and Education Reader.* New York: Teachers College Press.

Bennett, W. (1986). *What Works: Research About Teaching and Learning.* Washington, DC: US Department of Education.

Berliner, D., and Biddle, B. (1995). *The Manufactured Crisis: Myths, Fraud, and the Attack on America's Public Schools.* Reading, MA: Addison-Wesley Publishing.

Bigelow, W. (1990). Inside the classroom: social vision and critical pedagogy. *Teachers College Record,* 91(3), 437–448.

———. (1992). Inside the classroom: social vision and critical pedagogy. *Education Links,* 2(4), 351–357.

———. (1998). The human lives behind the labels, the global sweatshop, Nike and the race to the bottom. In W. Ayers, J. Hunt, and T. Quinn (eds.). *Teaching for Social Justice: A Democracy and Education Reader* (pp. 21–38). New York: Teachers College Press.

Bigelow, W., and Peterson, B. (2002). *Rethinking Globalization: Teaching for Social Justice in an Unjust World.* Milwaukee, WI: Rethinking Schools Press.

Carlson, D., and Apple, M. (Eds.). (1998). *Power/Knowledge/Pedagogy.* Boulder, Co.: Westview Press.

Cochran-Smith, M. (1991). Learning to teach against the grain. *Harvard Educational Review,* 61(3), 279–310.

Connell, R. (1993). *Schools and Social Justice.* Toronto: Our Schools/Our Selves Education Foundation.

———. (1994). Equity through education—directions for action. Canberra, Australia: Address to Australian Centre for Equity through Education.

———. (1998). Social change and curriculum futures. *Change: Transformations in Education,* 1(1), 84–90.

Cooper, K., and White, R. (Eds.). (2006). *The Practical Critical Educator: Critical Inquiry and Educational Practice.* Dordrecht, Netherlands: Springer.

Cox, R. (1980). Social forces, states and world orders' millennium. *Millennium: Journal of International Studies,* 10(2), 126–155.

Duncan-Andrade, J., and Morrell, E. (2008). *The Art of Critical Pedagogy.* New York: Peter Lang Publishing.

Feiman-Nemser, S., and Floden, R. (1986). The cultures of teaching. In M. Wittrock (ed.). *Third Handbook of Research on Teaching* (pp. 505–526). New York: Collier-Macmillan.

Garrison, D. (1991). Critical thinking and adult education: A conceptual model for developing critical thinking in adult learners. *International Journal of Lifelong Education,* 10(4), 287–303.

Ginsburg, M. (1988). Educators as workers and political actors in Britain and North America. *British Journal of Sociology of Education,* 9(3), 359–367.

Giroux, H. (1985a). Intellectual labour and pedagogical work: re-thinking the role of the teacher as intellectual. *Phenomenology and Pedagogy,* 3(1), 20–32.

———. (1985b). Critical pedagogy and the resisting intellectual. *Phenomenology and Pedagogy,* 3(2), 84–97.

———. (1988). *Teachers as Intellectuals: Toward a Critical Pedagogy of Learning.* Granby, MA: Bergin & Garvey.

Goodman, J. (1992). *Elementary Schooling for Critical Democracy.* Albany, NY: State University of New York Press.

Hinchey, P. (1997). *Finding Freedom in the Classroom: A Practical Introduction to Critical Theory.* New York: Peter Lang Publishing.

———. (2004). *Becoming a Critical Educator: Defining a Classroom Identity, Designing a Critical Pedagogy.* New York: Peter Lang Publishing.

Kincheloe, J. (2004). *Critical Pedagogy Primer.* New York: Peter Lang Publishing.

Kincheloe, J., and Steinberg, S. (Eds.). (1998). *Unauthorized Methods: Strategies for Critical Teaching.* New York & London: Routledge.

Kohl, H. (1983). Examining closely what we do. *Learning,* 12(1), 28–30.

McPeck, J. (1981). *Critical Thinking and Education.* New York: St. Martin's Press.

Popkewitz, T. (1987). *Critical Studies in Teacher Education: Its Folklore, Theory and Practice.* London & Philadelphia: Falmer Press.

Razack, S. (1993). Teaching activists for social change: coming to grips with questions of subjectivity and domination. *Canadian Journal for the Study of Adult Education,* 7(2), 65–78.

Shor, I. (1987). *Critical Teaching and Everyday Life.* Chicago: University of Chicago Press.

———. (1992). *Empowering Education: Critical Teaching for Social Change.* Chicago: University of Chicago Press.

Shor, I. (1996). *When Students Have Power: Negotiating Authority in a Critical Pedagogy.*. Chicago & London: University of Chicago Press.

Shor, I. (Ed.). (1987). *Freire for the Classroom.* Portsmouth, NH: Boynton/Cook.

Shor, I., and Pari, C. (Eds.). (1999). *Education is Politics: Critical Teaching Across Differences K-12.* Portsmouth, NH: Boynton/Cook.

Simon, R. (1992). *Teaching against the Grain: Texts for a Pedagogy of Possibility.* Amherst: Bergin & Garvey.

Smyth, J. (1991). *Teachers as Collaborative Learners: Challenging Dominant Forms of Supervision.* London: Open University Press.

———. (1993). A socially critical approach to teacher education. In T. Simpson (ed.). *Teacher Educators: Annual Handbook 1993* (pp. 153–165). Brisbane: Queensland University of Technology.

———. (1995a). Teachers' work and the labour process of teaching: central problematics in professional development. In T. Guskey and M. Huberman (eds.). *Professional Development in Education: New Paradigms and Practices* (pp. 69–91). New York: Teachers College Press.

———. (1995b). *Some Possible Candidates for Classroom Observation by Socially Critical Teachers and Colleagues.* Adelaide: Flinders Institute for the Study of Teaching.

———. (2000). Reclaiming social capital through critical teaching. *Elementary School Journal, 100*(5), 491–511.

Smyth, J., Hattam, R., and Lawson, M. (Eds.). (1998). *Schooling for a Fair Go.* Sydney: Federation Press.

Warham, S. (1993). Reflection on hegemony: towards a model of teacher competence. *Educational Studies,* 19(2), 205–217.

Wink, J. (1997). *Critical Pedagogy: Notes from the Real World.* New York: Longman.

Wood, G. (1988). Democracy and the curriculum. In L. Beyer and M. Apple (eds.). *The Curriculum: Problems, Politics and Possibilities* (pp. 166–190). Albany, NY: State University of New York Press.

———. (1990). Teachers as curriculum workers. In J. Sears and D. Marshall (eds.). *Teaching and Thinking About Curriculum* (pp. 97–110). New York: Teachers College Press.

Chapter 9

Empowering Education: Freire, Cynicism, and a Pedagogy of Action

Richard Van Heertum

Overview

If asked to describe Freire's work, some might invoke dialogue, praxis oppression, or possibly conscienization partly because Freire recognized that hope was at the very essence of human existence, the thing that kept the downtrodden waking up in the morning, the oppressed working, and the revolutionaries fighting against seemingly impossible odds. Indeed, hope may be as dangerous as knowledge to the power elites. What Freire recognized and struggled for his whole life was the belief that knowledge and hope could be brought together in a project of individual and collective emancipation from the sources of oppression and exploitation and toward a more just and equitable world.

In this chapter, I focus on Freire's notion of hope and unfinishedness in the struggle to combat the cynical undertow of contemporary neoliberal educational reforms. I begin with an analysis of cynicism and the ways education may work to foster it among the young. I then offer an overview of the ways in which Freire's ideas continue to resonate in the contemporary environment and how he can help inspire progressive educators to fight the multiple barriers to empowering students and teachers alike. I conclude by using Freire as the foundation to describe three approaches to pedagogy that can work to counteract cynicism and the exodus from hope. The first builds on the aesthetic education of German critical theorist Herbert Marcuse. The approach is useful throughout the curriculum as a way to counterbalance the rote memorization and drilling so endemic to No Child Left Behind (NCLB), exploring alternative affective responses to the surrounding world. The second is a form of civics education that involves student action and reflection, based on a four-part class on dolphins that I taught at an elementary school in Brooklyn. Civics education is largely absent in classrooms today and, following the lead of Dewey and Freire, I believe programs like this can teach kids the rudiments of democracy while sparking hope in the possibility of change. Finally, there is critical multiple literacy, which I argue is essential today with the growing influence of media culture on our lives and its close connection to neoliberal ideology and the consumer culture industry.

Throughout I argue that we must counteract the general tenor of NCLB and other neoliberal reforms that are making education about training alone and undermining the opportunities of millions of poor and minority children. I argue that neoliberalism not only spreads cynical views of knowledge and schooling but works to disempower teachers in the classroom and the spirit of public education in general, eschewing broader notions of education as a necessary component of a truly democratic society that reflects the interests of all citizens and works toward social justice and a reasonably equitable allocation of resources and opportunities.

The Roots of Cynicism

Has there been a collapse of political imagination in the West? Have we lost the ability to envision a world outside the strictures of conventional wisdom? And could this lead democracy to wilt under the force of collective neglect? The election of 2006 and the 2008 Barack Obama campaign certainly offer a spark of hope, challenging a conservative juggernaut that has dominated Washington politics for nearly thirty years, reestablishing widespread grassroots democratic participation and further solidifying new technologically enabled networks and public spheres from the fading vestiges of the past. Nonetheless, it is unclear how much change the election will precipitate, given a conservative transformation in the American political climate that has challenged many progressive gains of the twentieth century and altered the very nature of the relationship between the citizen and the state.

Today we appear trapped in the throes of an epoch where the "can be" is shut off by the "is" or the "should be." Retrenchment overruns dreaming, refinement reconstruction, and reform revolution. Where once there was imagination and struggle, we now appear trapped in an epoch of closure where we forego grand visions of the past like turn-of-the-century populism, the New Deal, Lyndon Johnson's "Great Society," the civil rights struggle or the 1960s countercultural revolution that once emboldened people to act. A silent consensus seems to disavow the government's role in solving social problems or improving our lives, accepting Thomas Paine's advice that "the government is best that governs least" without possessing any of his idealism or radical call to action. Today, many question the power of the state to do anything but protect us from terrorists or ensure our global economic competitiveness (Apple, 2001; Giroux, 2004), with few heeding Theodore Roosevelt's erstwhile argument that the federal government could stand as an "impartial tribunal, which would adjudicate between public good and private profit," or JFK's call to "ask not what your country can do for you, but what you can do for your country."

This is particularly true among the young, who have largely disengaged from active civic engagement and even voting—reestablishing what many believe is a meme of conformity not seen since the 1950s, though in a more diversified, ironic, and falsely rebellious sheen. Only 49 percent of those between eighteen and twenty-nine participated in the "most important election of our lifetime" in 2004, and the UCLA Freshman Survey found a decline in political interest, crashing from a high of 60 percent in 1966 to 34 percent in 2004 (Saenz et al., 2004). The Pew Research Center confirmed this trend, finding that the percentage of the young who registered a complete lack of interest in politics rose from 12 percent in 1987–1988 to 24 percent in 2002–2003, and the number "not too" or "not at all" interested in campaigns reaching an astounding 64 percent. Only 23 percent of those aged eighteen–twenty-nine even read a newspaper on a regular basis (*Online News Audience*, 2004). And this is compounded by findings from a 2005 John S. and James L. Knight Foundation study, which discovered that 33 percent of students thought the first amendment "went too far" in the rights it granted and over half thought newspapers should be censured (Yalof and Kautrich, 2005).

At the heart of this consensus is neoliberalism and its cynical instrumentalization of knowledge and exodus from hope. Neoliberalism is the dominant global paradigm today, founded as a project of market liberation, government retrenchment, and a dismantling of the social safety net. Its defining principle is that the iconographic "market" has the power to efficiently and effectively mediate the production and allocation of most social goods from consumer products to governance to medical care, energy, retirement funding, and education. Toward this end, it calls for privatization, diminished government oversight, reduced personal and corporate taxation, and, really, establishment of market ethics and rationality across all economic, political, and social institutions (Bourdieu, 1998; Giroux, 2004; Torres, 2005). In return, it promises equality of opportunity, democratic liberation, and meritocratic advancement, or as close to those ideals as is *possible*—and by maximizing their own self-interest, people optimize the outcome for all (Apple, 2001; Torres, 1998).

Neoliberalism thus redefines the role of the citizen in the state. No longer is she an autonomous architect of her reality, but a player in a market that serves as the ultimate arbiter of wants, needs, and desires. Citizens become "consumers" who contribute to society primarily through the work they do and the choices they make (Corner and Pels, 2003). The importance of the nation-state diminishes, and global markets and liberal democracy become accepted as universal, immutable truths (Castells, 1996; Hardt and Negri, 2000). As Giroux (2004) argues, neoliberalism thus "thrives on a culture of cynicism, insecurity and despair. Conscripts in a relentless campaign for personal responsibility, Americans are now convinced that they have little to hope for—and gain from—the government, nonprofit public spheres, democratic associations, public and higher education" (p. 105).

Neoliberalism is arguably cynical at its core and is thus a key battleground in attempting to reestablish democracy from the capitalist forces that have absconded from it. It sees humans as self-interested, greedy, and duplicitous, as do cynics. It implicitly distrusts government and most social institutions, as do cynics. It is founded on a belief in the immutability of present circumstance and the eternality of the status quo, as is cynicism. And it places almost blind faith in the unencumbered market, facilitated by cynics who provide no challenge to the increased encroachment of the market and corporations on every corner of the public and private spheres. Rather than believing in the institutions of democracy that are the fount of the original "American Dream" of equality and freedom, private interests, profit maximization, and their role in the forwarding of the economy are all that is left to believe in (Caldwell, 2006).

I believe there is truth in all of these claims. Cynicism appears to be described by three facets—distrust of human nature, distrust of social institutions, particularly governments, and lack of hope in the possibility of change. Its roots arguably lie in the economic, political, and social spheres, driven by the triumvirate of neoliberal ideology, political corruption and spectacle, and the postmodern condition of alienation, insecurity, atomization, fragmentation, and inauthenticity. Cynicism is intimately tied to the logic and discourse of neoliberalism, as previously discussed, particularly within education.

It is related to the current political landscape, where spectacle, manipulation, and dogma dominate campaigns, a corrupt, scandal-ridden government lacks accountability to anyone but the elite, and an economic consensus across party lines often allows social issues to trump economic interest and undermine democratic participation (Dean, 2004; Frank, 2004; Kellner, 2005). Finally, cynicism is a postmodern sensibility, where an atomized and fragmented public suffers from the implosion of the social ties, trust, and hope necessary for collective political engagement (Bauman, 1999; Bewes, 1997).

Education clearly plays a critical role in spreading cynicism, reinforcing hegemonic ideals, reproducing current power relations, and cutting off the channels for resistance and dissent. Gramsci, the first to systematically explore the power of education in reproducing the social order and prevailing rationality, followed by the work of Marcuse (1964), Bowles and Gintis (1976), and Paulo Willis (1981) among many, analyzed the profound role education plays in establishing a set of needs and desires that were then naturalized into universal, ahistorical truths.[1] These needs and desires were in line with domination and control and the maintenance of current power dynamics, cutting off the ability to think outside the prevailing rationality. More recently, a host of theorists have highlighted the antidemocratic social reproductive role of schools in breeding cynicism and a collapse of hope in the possibility of change (Apple, 2001; Aronowitz, 2001; Darder, 2002; McLaren, 2000). Giroux (2003) in particular highlights the way neoliberal reforms in schools focus predominantly on their economic role, forgoing the development of the whole student.

Yet how do schools actually teach cynicism? Education is one of the most powerful forces working to shape youths' social and political identities, together with the family, media, and religion. It does this in a number of ways, through pedagogical practices, curriculum, teachers and peers, and through the organization of the school and classroom. Schools have always served a central role in reproducing the social order, and today that includes the ethos of business, consumer culture, and neoliberal notions of historical immutability. NCLB and other neoliberal reform movements only amplify these concerns by demonstrating the close link between schools and the business community. This is coupled with attempts to extricate the political from education at all levels, redefining its goals and purposes in purely instrumental, economic terms.

While it is beyond the scope of this chapter to explore these trends in detail, I believe schools work to indirectly inculcate cynicism in four main ways. The first is through neoliberal educational reforms that instrumentalize knowledge and attempt to depoliticize schools and thus give knowledge the false sheen of neutrality.[2] Second is the relative absence of civics education in schools and the ways in which youths' imaginative and critical thinking skills are undermined by the strong focus on tests and back to basics education (Giroux, 2003; Tolo, 1999). Third are positivistic trends in educational research itself that tend to undermine efforts to redefine its purposes, goals, and values (Harding, 2004; Van Heertum, 2005). The fourth includes the numerous barriers to equality of opportunity along the lines of race and class, including deficit thinking, uncaring teachers, low expectations, funding differentials, tracking, and the resegregation of schools along racial and class lines (Kozol, 2005; Oakes, 1986; Orfield and Yun, 1999; Valencia, 1997; Valenzuela, 1999).

More critical and progressive approaches to pedagogy need to address each of these challenges if we hope to turn the tide in the struggle around what and how our children learn. In the interim, I believe it is essential that we recruit individual teachers and professors to begin the process of subverting the curriculum from within. Toward that end, I offer three strategies that can address the less tangible ways in which cynicism is spread: by opening student's minds to the surrounding world and ways to change it, by providing an empowering civics education that can inspire them toward political engagement, and by providing them with critical multiple literacy to highlight the effects of media (and education) on their lives and proffering them with the skills necessary to create their own alternative media, so central to the political landscape today (as displayed by, among others, Howard Dean, the Netroots, and Barack Obama).

Freire and the Centrality of Hope

While education and media play a central role in reproducing the social order, they also provide opportunities to rupture that order and offer alternative modes of social organization. Freire believed hope was essential to this project, facilitating an education that could truly challenge the injustices and inequalities of the past and present. His pedagogy thus starts from the position that we must help people recognize not only their oppressed situation, but their position as subjects in history with the power to change it.

Freire dedicated his final book, *Pedagogy of Freedom* (1998a), to helping teachers and students overcome the cynical fatalism at the heart of neoliberal ideology. He believed teachers needed to do more than make their students aware of inequality and suffering; they need to simultaneously empower them to believe that they could transform the world. Freire argued that we must embrace our "unfinishedness" in the world and recognize that change is the only constant in history. To him, the future was never preordained unless we accepted it as such. He thus argued that the "global tendency to accept the crucial implications of the New World Order as natural and inevitable," (1998a, p. 23) simply revealed the power of hegemony to spread, through education, the media, and discourse, the tenets of neoliberalism and their underlying logic of domination and control. Freire (1998b) believed that while we are conditioned, we are not determined, and are thus free to revolt against that conditioning:

> Our being in the world is much more than just "being." [It is] a "presence" that can reflect upon itself, that knows itself as presence, that can intervene, can transform, can speak of what it does, but that can also take stock of, compare, evaluate, give value to, decide, break with, and dream. (pp. 25–26).

By taking this position, he eviscerates the deterministic stance of neoliberalism, founded on a vision of reality dominated by extreme individualism, instrumental rationality, and, though unstated, subjects as essentially passive receptors of the events that surround them. Freire instead argues for a reality founded on dialogue in which students work in solidarity to first envision their surrounding reality and then work collectively to change it. Freire then argues for great vigilance in recognizing that teaching is not simply transmission of knowledge—whether it be hegemonic or counterhegemonic in nature. Teachers must respect the knowledge that students bring into the classroom, constantly question their own assumptions and techniques, and work to embrace cultural and ideological differences in order to open rather than close the student's mind. The progressive teacher must build on our collective "unfinishedness," showing students the profound power of social conditioning while respecting their autonomy and creative impulse to look at the world in ways different than progressives might find appropriate. This is fortified by the power of hope to embolden students to become excited about education, knowledge, and political action. Progressive educators can then spread hope by embodying it themselves, ensuring that their practice is self-confident, competent, generous, committed, humble, dialogical, caring, full of love, and able to effectively balance freedom and authority.

Ultimately, Freire can inspire progressive education primarily by returning to the foundation of his work and its focus on hope. In overcoming neoliberalism and its underlying fatalism and instrumentalization of knowledge, Freire's critique (1970) of banking education and call for problem-posing education as a way to open rather than close the minds of students is essential, as is his insistence on praxis-driven education that germinates from the everyday experiences of children. Freire's insistence (1998b) on the intimate connection between education and politics provides a solid foundation for challenging standardization in curriculum and testing, currently undermining the power knowledge has to emancipate people from ideological and material constraints. Freire's notion (1973) of limit-situations can also play a role here, providing teachers with the necessary background information to consider the barriers that stand in the way of their students' educational success and ways to overcome them. Likewise, Freire's consummate call for dialogue reminds us that teachers are the ultimate arbiters of what occurs in their classrooms, even as neoliberal reformers attempt to provide "teacher proof" curriculum that disempowers educators and close one of the last spaces where the creative and critical faculties can be cultivated. Finally is his belief in hope and the power that education has to remind students that they are indeed subjects in history who can work individually and collectively to reorient the world toward the needs of the many.

Aesthetic Education: Bringing Together Eros, Reason, and the Senses

Herbert Marcuse offers specific pedagogical strategies to work toward a more just, equitable, and humane society that can combat the increasingly firm stranglehold of capitalism on our wants, needs, and desires. In *Eros and Civilization* (1966), he offers mechanisms to approach a new sensibility where rationality and the senses come together allowing reason and happiness to converge. Toward this end, he calls for an aesthetic education that can break through instrumental and technological rationality.[3] This aesthetic education can work in concert with ideology critique and consciousness raising, serving as catalysts to reignite the belief that change is possible. In many ways, this approach comports with Freire's call to foreground affect and emotions as essential to learning. It also provides avenues for creativity and critical reflection in line with Freire's call to foster "epistemological curiosity" (Freire, 1998a).

The first aspect in the pedagogy is arts education, where production and appreciation become key components of learning. Marcuse (1972) argued that art offers an opportunity to step outside the dominant discourse and rationality to both deconstruct society and offer alternative dreams for children. Art is one form of the great refusal, a rejection of the discourse and rationality of an epoch—allowing one to consider alternative modes of thinking about and being in the world. As Ernst Bloch (1986) noted, gaining appreciation for art and a more critical view of popular culture can aid children in beginning to discern the traces of deeper libidinal desires that contemporary society fails to satisfy. Additionally, art offers an opportunity to awaken and foster the imagination, which has become increasingly verboten in contemporary schools and the larger society. Maxine Greene (1986) has advocated poetry and art for this very reason, seeing them as conduits for students to contemplate a different future. There is no reason not to go even further to incorporate music, media production, street art, and theater.[4] Many project-based learning advocates stress the ways that creativity and student autonomy can enhance learning, backing this up with solid empirical research (Blumenfeld et al., 1994; Thomas, 2000). Today, arts classes are among the first to be cut as schools struggle to meet state and federal standardized testing mandates, and one role of the progressive educator could be to both advocate their continued importance while incorporating arts education across the curriculum (Kozol, 2005).

Over the last few years, I have found an incredible array of arts projects undertaken by teachers to empower their students to explore the world around them and struggle toward change. One teacher I know in Los Angeles used film production in a South Central school as a method to provide a space for students to explore their surrounding in a critical way and create films that offered profound critique and channels of expression and inspiration to produce their own art and present it to their community. This is similar to the approach of Reach LA, an organization that combines media arts and technology with health education, teaching students media production techniques they can then use to create public service campaigns addressing HIV/AIDS, homophobia, racism, and other pervasive problems facing urban teenagers (see later). The partnerships forged between artists and students allows for the rearticulation in their own voices of pertinent social issues in the community. Several other teachers I know use spoken word as a method to bring emotions into the classroom and open a space for creativity, passion, catharsis, dialogue, and action. One of these teachers recently presented at a conference

I organized, mixing spoken word performance, pedagogical insights, and audience participation to show the power of art to transcend the limitations of traditional teaching and learning strategies. Finally are a series of media literacy projects that provide students of all ages with tools to explore the effects of media and consumer culture on their lives and produce their own alternative media that can challenge dominant modes of representation, exclusion, and alienation.

Sensual education is a second facet of Marcuse's aesthetic education. Like Dewey (1916) before him, Marcuse believed all the senses to be key in the learning process. He thus argued in *Counterrevolution and Revolt* and other works that we should incorporate the body more fully into learning, allowing students to explore the world in new ways. This will help shatter the false Cartesian logic of a separation between mind and body and open the door to a more immanent perspective on existence that escapes the dangers of transcendental thinking, where the present is sacrificed for an unknown future. In the past, music served as a powerful counter-hegemonic force and studying protest music from the past and present (like more radical hip-hop) at a more analytical level could introduce students to alternative ways of experiencing the world and escaping the mainstream music and visual culture so closely aligned with consumer culture rationality. Using other senses such as smell and touch could also be useful in challenging the strong proclivity toward vision so pervasive in Western culture, expanding the mind and allowing other ways to experience the world that surrounds us. Project-based learning again offers tools toward this end, by creating comprehensive curricula that allow a multitude of learning settings and strategies. As McLaren (1999), hooks (1994), and Shapiro (1998) among many have argued, teachers need to find ways to make the body present in the classroom, breaking the strong connection of desire, seeing, and cognition so prevalent today. This involves escaping the banking organization of the actual classroom and allowing students the space and time to more fully incorporate movement and action into learning.

A third element of the pedagogy is a return to nature. Marcuse (1972), unlike Freire, argues for a symbiotic relationship with nature in which we overcome a rationality that subordinates nature to human domination. This can be instrumental in overcoming the logic of humans dominating humans, by challenging its roots in the social and scientific domination of nature. If we can imbue children with an appreciation of nature as something we are part of, they may start to see the symbiotic relationship of all life. In this shift we reconnect with ecological and environmental movements and their belief that we are part of a world that offers us pleasure outside commodification. In education, this means getting students out into natural settings where they can explore their sensual relationships with nature outside the colonizing logic of science and classification. Poetry could be a good accompaniment here, providing an opportunity to explore the more spiritual relationship between children and the surrounding natural world. Environment as an Integrated Context for Learning (EIC) projects has been effective by moving in this direction.

A final aspect of an aesthetic education worth noting is Marcuse's efforts to follow the feminist movement in arguing for a pedagogy that escapes the patriarchal, andocentric and aggressive system of reason and attempts to embrace a sensibility more commonly affiliated with femininity. This admittedly essentialized view would involve a movement toward empathy, sensitivity, peace, caring, and nonexploitative practices—challenging the conservative, masculine skew of mainstream media and, often, education. Researchers such as Muller (2001), Parcel and Dufer (2001), and Noddings (2003) have shown the importance of a caring, nurturing environment to effective education, particularly for underperforming students (Eaker-Rich and Van Galen, 1996; Lyman, 2000). In this movement, another key idea from *Eros and Civilization* comes to the fore—which is the rechanneling of libidinal desires into an intersubjective, communitarian system of beliefs where a nonrepressive reality principle can come more into line with the pleasure principle. Groups like the Anti-Defamation League and the ACLU have already created programs in this vein, but it is then a question of incorporating them into the core curriculum or creating new curriculum while ensuring that the themes fit within some state or federal standards to allay administrative pressure and ensure progressive teachers stay in the classroom. It also involves something dear to Freire's heart, which is having faith and high expectations for all students and a shared commitment to their learning.

The Dolphin Project: Civics Education in Praxis

A 1999 report by the University of Texas LBJ School of Public Affairs found that American schools failed to provide students with even a rudimentary education in civic engagement and democratic participation (Tolo, 1999). The report found that while civic education was essential to a properly functioning and vibrant democracy, curriculum and standards in most states did not give adequate attention or resources to ensuring its inclusion. The report found that "students do not have the civic knowledge, the higher-order civic intellectual skills, and the civic disposition necessary to connect civic facts and concepts to the responsibility of citizenship." The report argues that lack of focus on civics education in professional development and pre-service teacher education, the focus on testing and accountability by administrators, and its absence in state and local standards all contribute to the problem.

A few years ago I undertook a project that attempted to place civics education back in the classroom through a praxis-driven pedagogy in line with Freire's notion of student-driven action research (Torres and Morrow, 2002). It involved a four-part class on dolphins for fifth graders in Brooklyn that combined marine biology, writing, dialogue, games, and political action. The idea of the project was to find ways to educate children on environmental issues while simultaneously providing them with a rudimentary education in civic engagement. The broader goal was to make this part of a larger project to inform the public about the killing of dolphins in Japan, using a strategy similar to one undertaken years earlier in the extremely effective dolphin-free tuna campaign. While the pilot project was successful, the larger project has yet to launch. But for me, the underlying idea resonated in a larger sense, as a model for engaging students in civic education through action that both allows them to recognize their role in our democracy and the part they can play in their local communities by building on the culturally specific knowledge they bring into the classroom.

The first session involved an interactive discussion on dolphins that brought up key themes and allowed connections to content standards. I also taught students about animal husbandry and engaged in an activity where they became dolphins themselves, trained to "jump" through a hula-hoop. It was a fun and engaging class that had the students excited about learning and their own knowledge. Class assignments that followed the first session involved a review of the material and homework assigned by the teacher to do independent research. The second class introduced the problem, in this case the killing of dolphins in Japan, together with a more scientific description of dolphins, their daily lives, and the ways in which eating dolphin and whale meat was a potential health hazard. In this class, a world-renowned marine biologist taught marine biology, public health, and scientific research techniques in an engaging and entertaining presentation. Class assignments included Internet research and an art project to conceptualize the killing of dolphins in drawings and paintings.

The third session then involved a visit to the New York Aquarium, where the students had the opportunity to meet a dolphin up close and one of the trainers who gave them a private demonstration, in addition to a general tour of the aquarium. At the end of the visit, we asked the students to write a letter to the Japanese premier and their congressperson, asking them to help stop the killing of dolphins for meat. This letter-writing assignment allowed them to work on their creative and persuasive writing skills and to tie writing to real-world problems The final class included a game of jeopardy to review the knowledge they had gained and a general lesson on how these activities related to political action in general, offering a model to engage in their own community to solve problems.

I sent the letters to the Japanese government and a local Congressman and arranged for a response from the latter. We also received local media coverage and extremely positive response to the program from the students, teachers, and administrators. While this project benefited from my association with the Wildlife Conservation Society in New York City, I believe it is a model teachers can effectively run in any classroom, strengthened if the issues relate to the community in which students live. The approach involves finding a problem important to the children, creating a curriculum that can address that problem in interesting ways while tying to standards, developing a plan of action that can reap some tangible positive outcome, and creating a bridge to a more generalized discussion of civic engagement.

The project then involves an initial dialogue to determine the topic (for older students) or a general discussion on topics to gauge student interest (among younger children). An introductory session or series of sessions then ensues, where the teacher raises the issue in general terms and creates a knowledge base. A second session or series of sessions then focuses on the problem or social issue associated with the generalized subject. In this session, an outside expert is ideal to keep student interest high and offer a level of legitimacy to the discussion. In the third session or series of sessions, students take some action to address the issue, hopefully in a hands-on manner that allows them to connect directly with the environment or problem. Finally is the fourth aspect, which involves dialogue and a more generalized discussion on the nature of political engagement and the power students themselves have to address social and political issues.

The framework allows for the incorporation of standards content, writing, reading, research, dialogue, and art. It connects learning to real-life issues and the culturally specific knowledge students bring into the classroom. The model follows the basic tenets of project-based learning and service learning while connecting education to civic engagement in a hands-on way that relates back to central tenets of Freire's pedagogy and the broader goals of empowering students and sparking hope.[5] I believe projects in this vein will be accepted by principals and parents, welcomed by communities, and embraced by students starved for curriculum that pushes the limits of their abilities, rather than curriculum based on low expectations, decontextualized knowledge, rote memorization, and incessant drilling.

Critical Media Literacy

Beyond education, the tentacles of media culture continue to spread outward across the social landscape, altering the very dynamics of how we live and perceive the world around us. A recent study from the Kaiser Family Foundation exemplified its broad reach, finding that "generation M" (those aged eight–eighteen) consume the equivalent of 8 ½ hours of media and technology each day, effectively occupying a third of their waking life (Rideout, Roberts, and Foehr, 2005). According to PBS's *Frontline* and the American Academy of Pediatrics, they are inundated with over 3,000 advertisements a day and watch 1,023 hours of television a year, outstripping school by over 100 hours. And Pew finds they rely heavily on television and the Internet to get political news, increasingly mock news programs like *The Daily Show* and *Colbert Report*. When videogames, movies, and music are added, it is clear that media exposure far exceeds time spent with family and friends, in school or any other activity, becoming the de facto in locus parenti.

Given this reach, I argue that cynicism is spread through not only education and discourse but the words, images, and actions of exemplars from Hollywood, television, and consumer culture. I posit that the cool, desirable, and heroic figures of movies and television often directly and indirectly promote the central tenets of neoliberalism and its underlying cynicism—offering a dour view of human nature, negative portrayals of politics and social institutions, and redeemer figures that save society outside democracy and lack of hope in the possibility of change. Media provides normative models of thought, behavior, and action (together with stereotypes of gender, race, class, ethnicity, sexuality, etc.) that kids often embrace and replicate. These include four themes that fit within a cynical framework: (1) Anti-intellectualism and apoliticism based on irony and cool detachment; (2) disrespect for authority and celebration of the market; (3) embracing of consumer culture and its focus on materialism, fame at any cost and subjectivity as a brand we sell to others, and; (4) an indirect embrace of passivity and political disengagement, driven by a culture of fear and desire for savior figures that can solve social problems and maintain order (Adorno and Horkheimer, 2002; Arnett, 2000; Bewes, 1997; Caldwell, 2006; Cappella, 1997; Chaloupka, 1999; Corner and Pels, 2003; Giroux, 2004; Goldfarb, 1991; Goldman, 1996; Kellner, 2005; Langstraat, 2002; Marcuse, 1964; Sloterdijk, 1987; Van Heertum, 2008).

Given the increasing influence of media culture, together with the realities of a more global, multicultural world, there have been increasing calls to alter education to meet the evolving needs and realities of the twenty-first century. This has only amplified the importance of critical media literacy and its role in assisting youth in navigating the increasingly complex world in which they live. Advocates for media literacy generally base their position on one of two arguments. The first is that students must be availed with the skills necessary to succeed in the new economy, and the second is recognition that media culture plays an increasingly important role in educating children.

Critical media literacy combines these approaches with a more critical aim and project of empowerment, "teaching students to be critical of media representations and discourses, but also stressing the importance of learning to use the media as modes of self-expression and social activism" (Kellner, 2001, p. 336). It advocates moving beyond the acquisition of skills alone and on to a dialectical engagement of the negative effects and positive emancipatory possibilities of media literacy. It includes a strong critique of mainstream approaches together with an alternative pedagogy and a political project for democratic social change, and it promotes the production of alternative counter-hegemonic media, embracing its creative potential and power to allow students to step outside the dominant discourse and rationality.

Len Masterman (1990) argues for the importance of *critical autonomy* in this project, giving students the freedom to judge for themselves the relative merits of alternative possibilities rather than being indoctrinated into a particular worldview—either from the right or left. This follows from Freire's desire to respect the autonomy of students and to avoid indoctrination. Yet critical media literacy becomes a more effective tool for empowerment and social justice if it goes beyond respecting student's autonomy to ensure that marginalized individuals and groups are given the opportunity to tell their stories and express their concerns. Kellner (2001) argues, "Technologies can be used as instruments of domination or liberation, of manipulation or social enlightenment, and it is up to the cultural producers and activist intellectuals of the present and future to determine which way the new technologies will be used and developed and whose interests they serve" (p. 337).

I believe feminist standpoint theory can enrich critical media literacy, empowering marginalized groups to have a say in this process and offering useful tools to deconstruct student and teacher positionality and a launching point for media production projects tied more closely to social critique and transformation (Harding, 2004; Van Heertum and Share, 2006). I thus argue that synthesizing critical media literacy with standpoint theory allows for a richer version of pedagogy that Jeff Share and I label *critical multiple literacy education* (Van Heertum and Share, 2006). Critical media literacy offers the foundational tools that allow children to critically reflect on media representations and political economy, the creative and critical skills to look beyond hegemonic discourse, and the empowerment to become agents of change for a more just, tolerant, and democratic world. Standpoint theory adds the centrality of positionality, voice and power to the pedagogy, advocating for beginning from the voices and experiences of the most oppressed groups, who may be in a better position to analyze society then those who benefit from the current order of things.

But what does a critical multiple literacy education look like? Building on the foundations laid down by the New London Group, it starts from embracing difference and a movement toward the cultivation of civic pluralism (Group, 1996). It does equip students with the skills necessary to succeed in the new global economy (computer and technological skills, financial and economic literacy, understanding of how to research and link information, etc.), but as part of a larger agenda to challenge the "new world order" (Kellner, 2002; Luke, 1997, 2000). It alters the nature of discourse away from deficiency theories and sees cultural and linguistic diversity as assets leading to improving social, political, and economic life.[6] This includes a reversal of current trends of discouraging bilingualism in school and challenging standardized curricula that have no relevance to many students' lives. At the same time, it takes a more dialectic view of media and technology that recognizes the power of instrumental and technological rationality to blanket the social justice agenda within the imperatives of the market. It advocates media production to cultivate creativity and critical-reflexivity, but always within a broader critique of technology and the larger society.

Following Freire (1970, 1998a,b) and his contemporaries, it would not be a set curriculum with prescribed teaching practices or universal curricular content. Instead, it would offer a basic theoretical foundation and critical pedagogical strategies, including the deconstruction of media texts, the creative production of media by children, and the teaching of skills in the use of various technologies necessary for success in contemporary society. Content would be culturally specific to the children, ensuring their interest and engagement with the material—but also transcending the particularity of their experience to see the world from disparate perspectives. It would also integrate open dialogue and mechanisms as key aspects of the pedagogy to allow children to participate democratically in the learning process.

One exemplar of this approach is the previously mentioned REACH LA. It was founded in 1992 by four women dedicated to combing the arts with a social justice agenda. Their mission was to build a working partnership between urban teenagers and artists where youth could creatively design the ways in which their communities addressed pertinent social issues. To this end, they use Augusto Boal's (1985) theater techniques and Paulo Freire's critical pedagogy, working to embody many of the aspects of a critical multiple literacy education—including combining media production and deconstruction with social justice issues and praxis toward social transformation.

Reviewing a subset of the videos produced between 2001 and 2004, I found that they consistently orient themselves toward the voices and experiences of the groups under study—generally incorporating extensive interviews together with popular media images and cinematic techniques. They show an openness to discussions of their identity as LGBT youth and the challenges they face in navigating life's travails. The videos often transcend the particularity of those lives though, providing deeper critiques of homophobia, mainstream media, and general social prejudice, and they tend to offer positive messages that empower others to learn from their experiences and struggle to redefine their realities.

One video, *We Love our Lesbian Daughters* (2004), captures the nature of the program. The filmmakers interview parents and lesbian girls, talking about the experience of coming out, of community and school prejudice, and of the importance of family support to their lives. The film stresses pride and acceptance and the importance of family as a potential support network. Another, *Surfacing*, powerfully engages questions of sexual abuse through the voices of a series of victims. The video talks about coming to voice, about gay coding in popular media, and about tools to overcome the abuse (through messages like "it's never the victim's fault," "all it takes is courage," and "speak out!!!"). Others explore "gender benders" (*Are You a Boy or a Girl*), heteronormative television imagery (*Profit of Hate*), and gay dating (*Gay Girl on the Party Line*).

According to Executive Director Martha Chono-Hesley, the organization works toward the creation of a safe space of respect and trust for students, where they can feel comfortable sharing their personal concerns and problems. The students and facilitators begin the program with creative writing exercises to connect with and critically reflect on their personal experiences and problems. Through a collective process, they share their stories, and discuss, critique, and support each other in "wrap sessions." The students then move into other areas of media production having established a higher level of trust and interconnectivity. The process also involves focus groups to collect community input and establish collaborations. The videos exemplify a high level of trust, generally including the students' own voices exploring their lives as gay youth and the lives of their family, friends, and fellow students. One participant explained his positive experience with the organization. "She [Chono-Hesley] sees us as artists . . . we take part of ourselves into our videos. If you could teach that in high school, that would be really cool . . . everybody's life has value in it."

Creating messages that challenge the discrimination and hypocrisy of the dominant discourse and making intimate connections to students' lives give this work critical and emancipatory elements missing from many programs. As students develop media literacy and production tools, they are simultaneously becoming more critical viewers who can create counter-hegemonic media that address personal issues of poverty, homophobia, and racism while connecting it to broader structural issues in their community and the larger world. REACH LA's approach to analyzing and producing media from subordinate positions and then "studying up" to reveal the larger oppressive social structures serves as a useful archetype to a critical multiple literacy education that can make students more critical, active, and hopeful citizens diligently engaging in the public sphere.

Conclusion

In general, I am arguing for the potent role teachers can play as public intellectuals working to restore hope and empowerment in children, drawing lines of connection and engagement that can alter the nature of education and politics in the future (Giroux, 1995; Pinar, 2001; Posner, 2001). In this process, I believe progressive educators need to build on the inspiration of Freire, recognizing four often overlooked attributes of an effective educator. These are hope, passion, joy, and love. I have spoken extensively about the need to offer hope to students, but a second aspect involves the teacher maintaining hope in their own projects. Working as a progressive or leftist is a process wrought with failure, and it is thus essential that teachers maintain hope even in the face of great obstacles and failures. This means finding satisfaction in small victories and recognizing the challenges we will face throughout our careers. Passion is also essential, as it is hard to inspire passion and hope in others if one does not possess it themselves. Teachers must reflect this passion in words and actions inside and outside the classroom. Joy help to strengthen this passion, offering students the opportunity to laugh and enjoy the learning process. This is a difficult task, but clearly helps combat the cynicism and disengagement so common in classrooms today. Finally, love for one's students is crucial. This love means believing in them even as they fail, offering a caring and nurturing environment and giving them the freedom to construct their own worldviews. Love cannot merely exist as a reflection of paternalistic attempts to help those less fortunate, it means a love that emboldens them to hope, dream, and act on their own.

In *Teaching to Transgress* (1994), bell hooks argued that "the classroom remains the most radical space of possibility in the academy . . . undermined by teachers and students alike who seek to use it as a platform for opportunistic concerns rather than as a place to learn" (p. 12). Paulo Freire, Henry Giroux, and a host of others have argued for a similar potential in K-12 and beyond, seeing schooling, and education more generally, as an opening for a revaluation of values. Neoliberal reforms undermine this dream, altering the nature of schooling to predominantly serve the role of training. Lost is the more holistic notion of education that involves the formation of active citizens that will contribute not only to the economy but the public good. Weakened is the notion that schools can create more tolerant and socially able adults that are well-adjusted and prepared to deal with their familial and social responsibilities. Largely disabled are efforts to cultivate the imagination, to critically engage with the world and see it in diverse ways, and to teach students the rudiments of democratic participation and civic engagement. And disavowed are efforts for education to serve the project of social transformation. While these pedagogical projects merely offer entryways within the interstices of the neoliberal machine, they provide a first step along the path to reclaiming the promise of public education as the very fount of democracy and equal opportunity for all citizens. Hope stands at the forefront of this daunting but essential project.

Notes

1. See Torres and Morrow for an alternative perspective on social reproduction theory (Torres and Morrow, 1995).
2. See chapter 3 of my dissertation *The Fate of Democracy in a Cynical Age: Education, Media and the Evolving Public Sphere* (2008) for a comprehensive analysis of the ways education is implicated in spreading cynicism. Also see the following for further arguments on cynicism: Bewes, 1997; Blackhurst and Foster, 2003; Caldwell, 2006; Cappella, 1997; Chaloupka, 1999; Cutler, 2005; Langstraat, 2002; Macedo, Dendrinos, and Gounari, 2003; Miles, 1997; Moore, 1996; Papastephanou, 2004.
3. Marcuse's aesthetic education takes much of its inspiration from Schiller's aesthetic education in *On the Aesthetic Education of Man* (1983). Marcuse's *An Essay on Liberation* (1969) and *Counterrevolution and Revolt* (1972) give further articulation of these ideas.
4. The techniques of Augusto Boal's theater (1985) of the oppressed can be quite useful in this sense, giving children a voice and the tools to struggle against their subordination.
5. This project is of course not unique. A recent book by Aaron Schultz, *Spectacular Things Happen along the Way* (2008), offers a wonderful example of another program in this vein. Also see collection of programs by Lee and Mann (Lee, 1997; Mann and Patrick, 2000).
6. See Valenzuela (1999) for further insights on deficiency theories, which place blame for minority underperformance on biological, cultural, and environmental factors rather than structural inequalities and barriers.

References

Adorno, T.W., and Horkheimer, M. (2002). *Dialectic of Enlightenment.* New York: Continuum.

Apple, M. (2001). *Educating the Right Way: Markets, Standards, God, and Inequality.* New York: Routledge Falmer.

Arnett, J.J. (2000). High hopes in a grim world. Emerging adults' views of their futures and "Generation X." *Youth & Society*, 31(3), 267–286.

Aronowitz, S. (2001). *The Knowledge Factory: Dismantling the Corporate University and Creating True Higher Learning.* Boston: Beacon Press.

Bauman, Z. (1999). *In Search of Politics.* New Haven: Yale University Press.

Bewes, T. (1997). *Cynicism and Postmodernity.* New York: Verso.

Blackhurst, A.E., and Foster, J. (2003). College students and citizenship: A comparison of civic attitudes and involvement in 1996 and 2000. *NASPA Journal*, 40(3), 153–174.

Bloch, E. (1986). *The Principles of Hope.* Cambridge, MA: MIT Press.

Blumenfeld, P., Krajcik, J., Marx, R., and Soloway, E. (1994). Lessons learned: How collaboration helped middle grade science teachers learn project-based instruction. *Elementary School Journal*, 94(5), 539–551.

Boal, A. (1985). *Theatre of the Oppressed.* New York: Theatre Communications Group.
Bourdieu, P. (1998, March). L'essence du neoliberalisme. *Le Monde Diplomatique.*
Bowles, S., and Gintis, H. (1976). *Schooling in Capitalist America: Educational Reform and Contradictions of Economic Life.* New York: Basic Books.
Caldwell, W.W. (2006). *Cynicism and the Evolution of the American Dream* (first edition). Washington, D.C.: Potomac Books, Inc.
Cappella, J.N. (1997). *Spiral of Cynicism: The Press and the Public Good.* New York: Oxford University Press.
Castells, M. (1996). *Volume 1: The Rise of the Network Society.* Oxford: Blackwell.
Chaloupka, W. (1999). *Everybody Knows: Cynicism in America.* Minnesota: University of Minnesota Press.
Corner, J., and Pels, D. (2003). *Media and the Restyling of Politics: Consumerism, Celebrity and Cynicism.* London: Sage.
Cutler, I. (2005). *Cynicism from Diogenes to Dilbert.* Jefferson, NC: McFarland & Company.
Darder, A. (2002). *Reinventing Paulo Freire: A Pedagogy of Love.* Cambridge, MA: Westview Press.
Dean, J. (2004). *Worse Than Watergate: The Secret Presidency of George W. Bush.* New York: Little, Brown and Company.
Dewey, J. (1916). *Democracy and Education.* New York: The Free Press.
Eaker-Rich, D., and Van Galen, J. (1996). *Caring in an Unjust World.* Albany: State University of New York Press.
Frank, T. (2004). *What's the matter with Kansas?* New York: Metropolitan Books.
Freire, P. (1970). *Pedagogy of the Oppressed.* New York: The Continuum International Publishing Group, Inc.
———. (1973). *Education as the Practice of Freedom: Education for Critical Consciousness.* New York: Continuum.
———. (1998a). *Pedagogy of Freedom.* Lanham, MD: Rowman & LittleField Publishers, Inc.
———. (1998b). *Politics and Education.* Los Angeles: UCLA Latin American Center Publications.
Giroux, H. (1995). Teachers as public intellectuals. In A.C. Ornstein (ed.). *Teaching: Theory into Practice* (p. 105). Boston: Allyn Bacon.
———. (2003). *Public Spaces, Private Lives: Democracy beyond 9/11.* Lanham, MY: Rowman & Littlefield.
———. (2004). *The Terror of Neoliberalism: Authoritarianism & the Eclipse of Democracy.* London: Paradigm Publishers.
Goldfarb, J.C. (1991). *The Cynical Society: The Culture of Politics and the Politics of Culture in American Life.* Chicago: University of Chicago Press.
Goldman, R. (1996). *Sign Wars: The Cluttered Landscape of Advertising.* New York: Guilford Press.

Greene, M. (1986). In search of a critical pedagogy. *Harvard Educational Review*, 56(4), 427–441.

Group, N.L. (1996). A pedagogy of multiliteracies: Designing social futures. *Harvard Educational Review*, 66(1), 60–92.

Harding, S. (2004). *The Feminist Standpoint Reader: Intellectual and Political Controversies.* New York: Routledge.

Hardt, M., and Negri, A. (2000). *Empire.* Cambridge, MA: Harvard University Press.

hooks, b. (1994). *Teaching to Transgress: Education as the Practice of Freedom.* New York: Routledge.

Kellner, D. (2001). *Media Culture: Cultural Studies, Identity and Politics between the Modern and the Postmodern.* New York: Routledge.

———. (2002). Technological revolution, multiple literacies, and the restructuring of education In I. Snyder (ed.). *Silicon Literacies: Communication, Innovation and Education in the Electronic Age* (p. 18). New York: Routledge.

———. (2005). *Media Spectacle and the Crisis of Democracy: Terrorism, War, and Election Battles* New York: Paradigm Publishers.

Kozol, J. (2005). *The Shame of the Nation: The Restoration of Apartheid Schooling in America.* New York: Three Rivers Press.

Langstraat, L. (2002). The point is there is no point: Miasmic cynicism and cultural studies composition. *JAC: A Journal of Composition Theory*, 22(2), 293–325.

Lee, L. (1997). *Civic Literacy, Service Learning, and Community Renewal* (No. EDOJC9704). California: Office of Educational Research and Improvement (ED), Washington, DC.

Luke, C. (1997). *Technological Literacy.* Melbourne: National Language and Literacy Institute: Adult Literacy Network.

———. (2000). Cyber-schooling and technological change: Mutliliteracies for new times. In B. Cope and M. Kalantzsis (eds.). *Multiliteracies: Literacy Learning and the Design of Social Futures* (p. 3). Austrialia: MacMillan.

Lyman, L. (2000). *How Do You Know You Care?* New York: Teachers College Press.

Macedo, D., Dendrinos, B., and Gounari, P. (2003). *The Hegemony of English.* Boulder, Colo.: Paradigm Publishers.

Mann, S.E., and Patrick, J.J.E. (2000). *Education for Civic Engagement in Democracy: Service Learning and Other Promising Practices.* Indiana: Office of Educational Research and Improvement (ED); Washington, DC: Corporation for National Service.

Marcuse, H. (1964). *One-Dimensional Man: Studies in the Ideology of Advanced Industrial Society.* Boston: Beacon Press.

———. (1966). *Eros and Civilization: A Philosophical Inquiry into Freud.* Boston: Beacon Press.

———. (1969). *An Essay on Liberation.* Boston: Beacon Press.

———. (1972). *Counterrevolution and Revolt.* Boston: Beacon Press.

Masterman, L. (1990). *Teaching the Media.* London: Routledge.
McLaren, P. (1999). *Schooling as a Ritual Performance: Toward a Political Economy of Educational Symbols and Gestures.* New York: Rowman and Littlefield.
———. (2000). *Che Guevara, Paulo Freire, and the Pedagogy of Revolution.* New York: Rowman & Littlefield Publishers, Inc.
Miles, P.G. (1997) *Generation X: A Social Movement toward Cynicism.* Master's thesis, University of Nevada, *DAI* 36–01:8.
Moore, M.P. (1996). From a government of the people, to a people of the government: Irony as rhetorical strategy in the presidential campaigns. *Quarterly Journal of Speech*, 82(1), 22–37.
Muller, C. (2001). The role of caring in the teacher-student relationship for at-risk students. *Sociological Inquiry*, 71(2), 241–255.
Noddings, N. (2003). *Caring: A Feminine Approach to Ethics and Moral Education.* Berkeley: University of California Press.
Oakes, J. (1986). *Keeping Track: How Schools Structure Inequality.* New Haven, CN: Yale University Press.
Online News Audience Larger, More Diverse: News Audience Increasingly Politicized. (2004). Washington, DC.
Orfield, G., and Yun, J. (1999). *Resegregation in American Schools: The Civil Rights Project.* Harvard University.
Papastephanou, M. (2004). Educational critique, critical thinking and the critical philosophical traditions. *Journal of Philosophy of Education*, 38(3), 369–378.
Parcel, T., and Dufer, M. (2001). Capital at home and at school: Effects on child social development. *Journal of Marriage and the Family*, 63, 32–47.
Pinar, W. (2001). The researcher as bricoleur: The teacher as public intellectual. *Qualitative Inquiry*, 7(6), 696–700.
Posner, R. (2001). *Public Intellectuals: A Study of Decline.* Cambridge: Harvard University Press.
Rideout, V., Roberts, D., and Foehr, U. (2005). *Generation M: Media in the Lives of 8 to 18 Year-Olds.* Kaiser Family Foundation.
Saenz, V., Hurtado, S., Astin, A., Vogelgesang, L., and Sax, L. (2004). *Trends in Political Attitudes and Voting Behavior among College Freshmen and Early Career College Graduates: What Issues Could Drive this Election?* Los Angeles: CA.
Schiller, F. (1983). *On the Aesthetic Education of Man.* Oxford: Oxford University Press.
Schultz, B. (2008). *Spectacular Things Happen Along the Way: Lessons from an Urban Classroom.* New York: Teachers College Press.
Shapiro, S. (1998). *Pedagogy and the Politics of the Body: A Critical Praxis.* New York: Routledge.
Sloterdijk, P. (1987). *Critique of Cynical Reason.* Minneapolis: University of Minnesota Press.
Thomas, J. (2000). *A Review of Research on Project-Based Learning.* Reported completed for The Autodesk Foundation, March 2000.

Tolo, K. (1999). *The Civic Education of American Youth: From State Policies to School District Practices.* University of Texas at Austin.

Torres, C.A. (1998). *Democracy, Education and Multiculturalism: Dilemmas of Citizenship in a Global World.* New York: Rowman & Littlefield Publishers, Inc.

———. (2005). The NCLB: A brainchild of neoliberalism and American politics. *New Politics*, X(2), 38.

Torres, C.A., and Morrow, R. (1995). *Social Theory and Education: A Critique of Theories of Social and Cultural Reproduction.* Albany, NY: State University of New York Press.

———. (2002). *Reading Freire & Habermas: Critical Pedagogy and Transformative Social Change.* New York: Teachers College Press.

Valencia, R. (1997). *The Evolution of Deficit Thinking: Educational Thought and Practice.* Washington, DC: The Falmer Press.

Valenzuela, A. (1999). *Subtractive Schooling: US-Mexican Youth and Political Caring.* New York: State University of New York Press.

Van Heertum, R. (2005). How objective is objectivity? A critique of current trends in educational research. *Interactions: UCLA Journal of Education and Information Studies*, 1(2). Accessed on June 12, 2008 at repositories.cdlib.org/gseis/interactions/vol1/iss2/art5.

———. (2008). *The Fate of Democracy in a Cynical Age: Education, Media and the Evolving Public Sphere.* Los Angeles: University of California Los Angeles.

Van Heertum, R., and Share, J. (2006). Connecting power, voice and critique: A new direction for multiple literacy education. *McGill Journal of Education*, 41(3), 249–266.

Willis, P. (1981). *Learning to Labor.* New York: Columbia University Press.

Yalof, D., and Kautrich, K. (2005). *The Future of the First Amendment.* . Muncia, IN: John S. and James L. Knight Foundation.

Chapter 10

Teachers Matter... Don't They? Placing Teachers and Their Work in the Global Knowledge Economy

Susan L. Robertson

Introduction

In the conclusion to my book, *A Class Act: Changing Teachers' Work, the State and Globalisation* (2000), I argued:

> If history can be read forward... there is a simple lesson to be learned about teachers... the conditions associated with fast capitalism, the rise of the competitive contractual state and the tendency toward individualism and "doing well"... will have created new fissures and progressively fragmented teachers as a unified category of workers. (Pp. 211–213)

Reading history forward does not, of course, mean the future is determined for us. Far from it! However, eight years have passed since drawing that conclusion, and a great deal has happened in the world. Arguably the most significant development, at least for the purposes of this chapter—which broadly is intended to place teachers and their work in the context of globalization—is the emergence of a new, very powerful, discursive imaginary—the assertion that we now live in, or are moving toward, a knowledge-based economy.

The focus on knowledge as the key motor for the economy, on how to create, distribute, and manage it, has placed education at the center of policy and politics. Earlier versions of human capital theory have been invigorated by new growth theorists who argue that it is not just more education that matters, but kinds of education experiences that foster innovative aptitudes (Romer, 2007), while popular intellectuals, such as Richard Florida (2002, 2005) have promoted concepts like the creative class as the basis for producing competitive economies and cities. Among policy makers, there is now intense interest in developing creativity as a basis for invention and innovation, with the suggestion that new, more active, child-centered pedagogies are desirable, and should be promoted in schools. It is against this backdrop we are told "teachers matter" (OECD, 2006) and that high quality educational provision is now more critical than before.

This might suggest a move away from almost two decades of teacher blaming, audits, and managerialism that has accompanied much of the restructuring of education around the world toward a revaluing of teachers. However, in this chapter, I will be examining a number of projects underway that, if realized, have the capacity to generate profound changes to education and to teachers' work. At the heart of these projects "translating" the knowledge economy discourse into new institutional structures and material practices is the view that the education system must be radically transformed to secure the future. These projects include the "modernization of the school," "personalization" of learning, the "scientization" of teachers' knowledge, the "biologization/neurologization" of the learner, and the commodification of schooling. If I am right in my prognosis, then these developments can be read as a rupture in the grammar of schooling (Dale, 2008). In combination these projects, if realized, would lay the groundwork for a very different kind of "education laboring" for teachers and learners. They also raise fundamental questions about what kind of learner and what kind of society is being constituted. The chapter is developed in three parts. I begin with some reflections on the globalization of neoliberalism throughout the 1980s and 1990s and its consequences for teachers' work. I then turn to an examination of the knowledge economy discourses promoted particularly by the international agencies from the late 1990s. This master narrative has gained sufficient traction in state policy circles for it to legitimize a newer, deeper round of institutional innovation/transformation, including education. In the third part of the chapter I examine, briefly, four projects intended to advance this renovation and recalibration of education to constitute a knowledge-based economy. In the final conclusion I stand back and review the implications of these developments for contemporary societies more generally, and for teachers as laboring class in particular.

Globalizing Neoliberal Projects and Teachers' Work

To date, most teachers' experiences of globalization have come in the form of the globalization of neoliberalism that, according to Santos (Dale and Robertson, 2004), is a particularism that has secured for itself hegemonic status. Neoliberalism is a theory of political and economic practice that proposes human's well-being is best advanced by liberating their individual entrepreneurial freedoms and skills within an institutional framework characterized by private property rights, free markets, and free trade (Friedman, 1962).

In response to a global crisis of capitalism that shocked the world economies in the early 1970s, from the early 1980s onward governments around the world wittingly embraced neoliberalism as an alternative to ethical liberalism (Keynesianism), or unwillingly had this ideology imposed upon them largely as a result of IMF/World Bank Structural Adjustment Programmes (Robertson et al., 2007). The outcome was the radical transformation of the social fabric of societies around the globe. These transformations led David Harvey (2005) in his brief history of neoliberalism to observe: "Future historians may well look upon the years 1978–80 as a revolutionary turning point in the world's social and economic history" (p. 1).

Three central principles featured in neoliberal informed restructuring: deregulation, competitiveness, and privatization (Cox, 1996, p. 31). The first, deregulation, refers to the removal of the state from a substantive role in the economy, except as a guarantor of the free movement of capital and profits. The second, competitiveness, refers to the justification for dismantling existing political and economic structures and constructing new, more market-friendly ones. The third, privatization, describes the sale of government businesses, agencies, or services to private owners, where accountability for efficiency is to profit-oriented shareholders. Through prising open the growing fissures in the postwar class compromise and hastening its demise, neoliberals and their allies "re-leveled" and "re-bordered" the playing field, putting into place a set of rules that directed the steady flow of class assets (cultural, economic, social) upward toward the ruling classes (Robertson et al., 2007).

Education systems—particularly the school sector—were shielded from the full force and impact of this agenda. For instance, quasi rather than complete market principles were introduced into the schooling sector to break "provider capture" and open up the possibility for "choice" and education consumerism (Ball, 2002). This altered existing patterns of postwar redistribution and social mobility against the interests of the working and middle classes (Robertson and Lauder, 2001).

Drawing on discourses (and in some places reality) of "crisis," neoliberal political projects were mobilized by international agencies, transnational firms, and governments across national state spaces. These projects set about challenging and changing the architecture of schooling; its *mandate* (what it is that the education system should do), *capacity* (the means through which the mandate can be realized, e.g., fiscal and human resources), and mechanisms of *governance* of the education sector (i.e., the means for coordinating the system) (Robertson, 2000).

In terms of the "mandate" for education, the economy was prioritized above all else. Education systems were tasked with developing efficient workers for a competitive national economy, while teachers were to demonstrate through national (SATS) and global (e.g., PISA, TIMMS) systems of indicators they had taught their young charges "well." Regarding "capacity," there was an overall reduction in financing in the public sector more generally and in education. In general (aside from Latin America—whose expenditures on education rose from 3 to just under 4.5 percent) most regions experienced an overall decline in education expenditures as a percentage of GDP (ILO, 2004, p. 47). Education providers were pressured to use funds more efficiently and encouraged to seek additional sources of funding from households and the business sector (Robertson et al., 2007, pp. 43–47).

In sketching out the broad features of the political project, I am not intending to suggest that neoliberal projects, policies, and practices in education were implemented in the same way, at the same time, with the same effect, across national state spaces. The particular constellation of institutions and social forces in a formation will mediate the discourses and projects of actors. However, by anyone's reckoning, the advance of neoliberal projects has dramatically altered the social fabric and social relations of many societies around the globe.

In looking at the impact of neoliberalism on teachers' workplaces and conditions of laboring it is possible to detect affects on teachers' work, status, and market situations. A major report into teachers' work in twenty-five countries by the OECD (2005) provides some insights into the affects. Half of the countries in the OECD study reported problems of teacher shortages. Evidence suggests that shortages are the result of deteriorating conditions of work, such as heavy workload, lack of resources and support, pupil behavior, and ongoing government reforms creating a stressful work environment. However, where salaries are high, there are few problems (p. 74). In countries where there were shortages (Sweden, Finland, and Belgium), more than 10 percent of teachers appointed to cover classes were not fully qualified (i.e., having a degree in the area they are teaching in). In the United States, at least 20 percent of the teachers appointed to cover classes did not have a proper qualification. The problem of teacher shortages is more acute in secondary schools, and in low-income, low-achieving urban areas (pp. 49–50). Teacher attrition is also problematic, particularly in the United States, United Kingdom, Sweden, and Israel where it is above 6 percent, whilst there were only a small number of countries (Italy, Japan, and Korea—see p. 173) where teacher attrition was less than 3 percent. Attrition reflects the exit from the profession of early career teachers. These rates are: higher in secondary compared with primary schools (p. 176); amongst better rather than less well-qualified teachers; and amongst teachers working in more disadvantaged rather than advantaged schools. Overall, while teachers' salaries in real terms increased in almost all of the OECD countries covered in this study, in comparison to other occupations, teachers' salaries have fallen further behind. In general, since the early 1990s, teachers feel they have low status and little public respect. It is paradoxical then that while the knowledge economy discourse now places quality learning at the center of policy makers' agendas, one effect of more than two decades of neoliberal policies and programmers in education has been to so seriously erode teachers' working conditions that it has undermined teaching as a profession. This has not only placed teachers and the teaching profession in a situation where they are more vulnerable to critique but it has opened the space for the advance of a more radical "modernizing the school" agenda at national and international levels. This new economic imaginary draws its legitimacy from the claim that we are living in a knowledge-based economy, in turn necessitating an even more radical set of reforms to education than we have seen to date. Despite more than two decades of restructuring, education, it is argued, is a creature of the industrial age that continues to promote a "one size fits all" pedagogy and curriculum. In the following section I look more closely at the history of the knowledge economy argument in order to reveal both its politics and also the actors that are involved in this project.

The Knowledge-Based Economy Master Narrative

As I noted earlier, since the late 1990s "knowledge-economy" discourse has dominated talk in political and policy circles. Policy statements from the multilateral agencies, firms, and national governments of all persuasions assert that "we live in a knowledge-based economy" (cf. Blair, 2000; European Council, 2000; OECD, 1996; World Bank, 2003). It is important at this point to note the hortatory character of this claim—the insistence that a new ontological order has emerged. However, as we will see, "the knowledge-based economy" does not exist a priori. Like all economies "the knowledge-based economy" is constructed. It is a fluid and dynamic entity; an evolving outcome of ideational, representational, material, and institutional discourses, practices, and struggles.

The idea of a knowledge-based economy has its roots in work developed by a group of 1960s intellectuals, futurologists, and information economists, including Fritz Machlup (1962), Peter Drucker (1969), and the well-known Daniel Bell (1973). These writers argued that societies were in transition to becoming knowledge-based. Their thesis, regarded as highly speculative at the time, was later added to by Manuel Castells (1996, 1999) and his theory of the emergence of a network society. A core argument in this body of work is that information/knowledge is now a new factor in production.

The OECD was heavily influenced by these ideas. During the 1970s, the OECD took on board the idea of an "information society" (Mattelart, 2003, p. 113). It also enlisted the expertise of a range of economists concerned with mapping and measuring information. The concept of a *knowledge-based* economy was added in the 1990s, and reflected the contribution of economists, such Dominic Foray (2000) (that it was knowledge and not information that was important, and that economic growth was the result of the distribution and use of knowledge), Bengt-ake Lundvall (1996) (focused on processes of learning in firms), and new growth theorist Paul Romer (2007) (economic growth occurs when people take resources and rearrange them in ways that are more valuable).

The OECD then moved toward developing sets of indicators to both measure and guide national state's development toward a knowledge-based economy. The effect of producing statistics to measure the knowledge-based economy (KBE) in turn began to stabilize and materialize the idea of a knowledge-based economy around four pillars that the OECD and other international agencies and national actors were encouraged to agree upon: "innovation," "new technologies," "human capital," and "enterprise dynamics" (see Robertson, 2009, for a fuller explanation). These four pillars were also taken up in the World Bank's *Knowledge for Development* program launched in 1996. At the heart of the OECD's version of the knowledge economy is the idea that knowledge has value. As Bell (1973) put it:

> Knowledge is that which is objectively known, an *intellectual property*, attached to a name or group of names and certified by copyright, or some other form of social recognition (e.g. publication)... It is subject to a judgment by the market, by administrative or political decisions of superiors, or by the peers as the worth of the result, and as to its claim on social resources, where such claims are made. In this sense, knowledge is part of the social overhead investment of society, it is a coherent statement, presented in a book, article, or even a computer program, written down or recorded at some point for transmission, and subject to some rough count. (P. 176; italics in the original)

So, why the interest in the idea of a knowledge-based economy? We can begin to make sense of this if we set it against the crisis of capitalism in the early 1970s and the subsequent search for solutions to underpin the next long wave of accumulation. As we have seen already with the neoliberal project that drove the restructuring, crises are path-breaking and path-shaping moments. Crises also require both semiotic and strategic innovation.

However, while through the 1980s and 1990s neoliberal political theory provided the means to unpick old institutional structures and embed the basic architecture of market liberalism, the collapse of the Washington Consensus, the leakiness of neoliberal projects, and the global struggles around the WTO resulted in a series of renovations—Third Way politics, the Post Washington Consensus, and so on. Strategically, neoliberalism as an economy imaginary was not adequate to power forward and stabilize a new social formation. This is because the emergence and consolidation of a new economic regime is dependent upon *more* than changes in the economy: "It also depends critically on institutional innovation intended to reorganize an entire social formation and the exercise of political, intellectual and moral leadership" (Jessop, 2004, p. 166). This requires an economic imaginary that has considerable resonance, plausibility, flexibility, and interpretability. It must be one that also:

> enables the rethinking of social, material and spatio-temporal relations among economic and extra-economic activities, institutions, and systems and their encompassing civil society through proposing visions, projects, programmes and policies. And, to be effective, it must, together with associated state projects and hegemonic visions, be capable of translation into a specific set of material, social and spatio-temporal fixes that

> jointly underpin a relative structures coherence to support continued accumulation. (P. 116)

Through the 1990s, with steerage from dominant nations, regions, and agencies, such as the US, EC, WTO, OECD, and World Bank, the idea of a knowledge-based economy was promoted to eventually emerge as a powerful master economic narrative in many accumulation strategies, state strategies, and hegemonic visions around the world. And while it corresponds in significant ways to changes in technologies, labor processes, and forms of enterprise, as we have seen, it emerged out of the field of other possible contenders, including ideas such as the network society and informational age, and so on. The idea of "knowledge" is particularly potent in this discourse, as it is able to articulate with progressive left as well as right projects. Who can be against knowledge? It also articulates with both human capital and new growth theory, with their interest in the basis of economic growth and competitiveness. However, if we look more closely, the OECD and World Bank's approach is deeply inflected with Western-centered mercantilism (Jessop, 2004). This more neoliberal version of the knowledge-based economy seeks to deepen and widen its grasp space by presiding over an extension of intellectual property rights, establishing institutions to ensure that value is returned across borders (Robertson, 2009), privileging knowledge creation/venture capital initiatives, and developing of creative/innovative subjects for capital accumulation.

Given the central role of education in social reproduction and cultural production, it is hardly surprising that education systems around the globe would *again* be scrutinized more closely. Education systems are important (though not exclusive) sites for the production of knowledgeable subjects. It would be important, therefore, to realize a knowledge-based economy for education to be renovated in ways that would enable this new kind of self/worker/citizen to be constituted. An economy driven by constant innovation would require a rather different kind of self—one that actively produced new knowledge (and potential products and markets) through processes of assembling and reassembling knowledges. However, education systems have also increasingly been viewed as sites for profit-making. Until recently, education systems had been protected from the intrusion of capital by discourses of public good, public service, and human rights. However, in knowledge-based economies, where knowledge services have a value, it is also a logical move to bring education into the economy as a services sector in its own right. This requires the state to lose its monopoly hold over education and enable new players in. These two related moves have opened up education to a range of projects intended to re/construct the sector, its pedagogy, and subjectivities.

Translating and Constituting the Knowledge-Based Economy

Much of this problem specification agenda setting for the radical reorganization of education has come from the international agencies (OECD, WTO, WB), transnational firms (Microsoft, Sylvan Learning Systems), and think-tanks (such as Demos, Futurelab) (Robertson, 2005; Robertson and Dale, 2009). At a range of scales, projects are now translating, materializing, and constituting the master narrative of the knowledge-based economy. All have profound implications for the organization of teachers' work.

Modernizing the School for the Twenty-first Century

Work on the future of schooling was begun by the OECD with its *Schooling For Tomorrow* program (2000). The need for the program was justified on two grounds: the short-term basis of national policy-making and practice in the face of increasing complexity and change; and the fragmented and unscientific nature of education's own knowledge base. In order to focus attention on problems in the contemporary school sector, the OECD proposed a *Schooling for Tomorrow Toolbox* (OECD, 2000) aimed at identifying ways of enhancing decision-making at national and sub-national levels. Six scenarios are developed intended to challenge policy makers and practitioners to visualize desirable futures for schooling and how these might be achieved. Education leaders were encouraged to proactively influence their wider environment, redesign the way that organizations work, and shape their own country's futures based on national and global trends.

Three pairs of scenarios were developed in the "toolbox"—all possible responses to the problems of learning for the knowledge economy. These are: maintaining the "status quo" (schools as outdated bureaucracies), "re-schooling" (reorganizing to prioritize school as learning organization), and "de-schooling" (school as market of market network). The overall negative orientation to the status quo scenario as a description of the current organization of schooling was meant to convey the view that it cannot offer an adequate vision and orientation to the future. Both re-schooling and de-schooling were then selected as possible ways forward. Both privilege the learner above teachers, and new forms of governance over state monopolies, as the means of realizing knowledge-based economies. The OECD's preferred position tended toward the re-schooling scenario, with schools continuing to sit inside a web of state and private sector provision rather than a full-blown market model.

In its first major foray into education policy for secondary schools, the World Bank's 2003 *Lifelong Learning for a Global Knowledge Economy* (directed at developing countries), also tackles the need for the radical transformation of schooling. It reinforced Bell's views outlined earlier that "a knowledge-based economy relies on ideas rather than physical abilities and the application of technology rather than the transformation of raw materials or the exploitation of cheap labor . . . The global knowledge economy is transforming the demands of the labor market throughout the world" (World Bank, 2003, p. 161). The Bank then argues that the global knowledge economy "is also placing new demands on citizens who need new skills and knowledge to be able to function in their day-to-day lives. Equipping people to deal with these demands requires a new model of education and training, a model of lifelong learning" (ibid.).

In the report the Bank contrasts current education systems (status quo) with a "lifelong learning" approach. Current systems of education are argued to be teacher dominated, test-based, and focused upon rote learning. A lifelong learning model, by contrast, is based on "doing"; it would be pupil-driven and personalized, with individual learning plans. Teachers are viewed as impediments, imposing facts on students. Teachers should be guides and mediators. Space is also made for technologies to become knowledge-based tutors (p. 38). The prioritization of technologies and the Bank's commitment to public-private partnerships creates an entry point for transnational firms to enter into the education sector countries. The imagined school for the future for the World Bank is captured by the de-schooling scenario—with new technologies and the for-profit sector playing a significant role in the provision of learning.

More recently the European Commission (2007a) has also embraced the "modernizing the school" agenda as a means for realizing its own competitiveness agenda (EC, 2007b). This is a radical and controversial move given that schools are constitutionally protected by the principle of subsidiarity and are therefore part of national state space. Despite political sensitivities, the EC has pressed ahead, and is inviting member states to discuss the agenda at its November 2007 ministerial meeting in Lisbon, Portugal. The EC's working paper for discussion by member states reflects many of the same issues as the OECD and World Bank reports: the importance of education to develop the stock of human capital (p. 3); the need to modernize the education system to ensure the development of individual creativity; "the ability to think laterally, transversal skills and adaptability . . . rather than specific bodies of knowledge" (p. 5). The EC also notes that the persistence of social inequalities limits the success of education policies in ensuring successful learning for "young Europeans" (p. 9). In all, this is a less radical intervention by contrast with the OECD and World Bank. Its focus is on identifying the problems and issues facing member states in generating a competitive and cohesive Europe. However, in the conclusion, the EC points out that "the institution of the school cannot remain static if it is to serve as a foundation for lifelong learning" (2007a, p. 11). Member states are invited into proposing solutions that might enable them to modernize their systems. This more tentative solution-seeking approach is a consequence of the political reality facing the Commission in advancing its vision, project, and strategies at the European scale.

The "Scientization" of Teachers' Knowledge

A second strategic project area is the teacher. The concern is not with the wider conditions under which teachers work but the nature of teachers' knowledge. David Hargreaves' arguments (2001) have been very influential in OECD circles. He has also been very influential in the United Kingdom through his stewardship of key government agencies. Hargreaves argues teachers do not possess a body of codified scientific knowledge around teaching and learning. Rather, teachers work in individualized settings and acquire their knowledge through trial and error. Their knowledge is thus personal rather than collective, tacit rather than explicit, and subject/content based rather than process based. Two problems are identified here (OECD, 2001). The first is that teachers do not build up a body of evidence and use that evidence to inform their own practice. The OECD has kept the issue alive by running a series of conferences and workshops exploring how research evidence can be better used by teachers to inform teaching and learning (OECD, 2007). It has also created space for discussions on the kinds of institutions (such as completing reviews of research on areas like information and communication technologies or ICTs and learning) that might synthesize knowledge in ways useful to teachers. However, the tendency has been to generate a simplistic "what works"—or *x causes y* approach (supported by evidence from random field trials if possible), rather than a more context sensitive "what works for whom, under what circumstances, with what outcomes" approach, where complexity and contingency in social settings is taken into account.

The second approach derives from the influential work of Gibbons and colleagues; that content/discipline-based knowledge (Gibbons *et al.*, 1994, call this Mode 1 knowledge) is less important than process and trans-disciplinary knowledge (Mode 2 knowledge) in a knowledge-based economy. Drawing upon these kinds of arguments, the OECD (2001) claims that: "Teachers . . . now need to teach students to learn how to learn . . ." and that "this requires the production and application of new pedagogic knowledge on a huge scale" (p. 71). They add:

> The creation and application of professional knowledge on the scale and in the time-frame demanded by "schooling for tomorrow" makes demands at the individual and the system levels. At the level of the individual teacher, there needs to be a psychological transition from working and learning alone with a belief that knowledge production belongs to others, to a radically different self-conception which, in conformity with interactive models, sees the production of knowledge with colleagues as a natural part of teachers' professional work. At the system level ways have to be found to bring teachers together in such an activity. (Ibid.)

While crude forms of the scientization of teachers' work, particularly those around "evidence-based practice," are viewed by teachers with skepticism and resistance, many teachers have been motivated to work in more collaborative, interactive ways and embraced opportunities that enable this. They have also been keen to take advantage of opportunities offered by governments to develop partnerships with universities to coproduce—though research—knowledge about improving learning. These developments are having a positive effect on teachers' work and suggest that projects of this kind will "fix" new pedagogical practices.

Personalization and the "Prosumer"

A third project being advanced is personalized learning. This strategy is a response to the problem of "learning how to learn" and has been finessed by the OECD, the U.K. Department for Education and Skills, and U.K.-based think-tank Demos. Personalization is a key strategy within the social policy sector more generally (Ferguson, 2007) to produce "active citizenship" (Jenson and Saint-Martin, 2006). It challenges current ambitions for reform. That is, the OECD argues that current visions/practices do not have the future (post-industrial) reality in its sights. Personalization sets out to generate a new social architecture and subjectivity through recalibrating the social policy/program/consumption mix. Personalization also replaces words like consumerism in an effort to create an effect of distance between the earlier neoliberal project and the knowledge economy master narrative, though as we will see they are tightly linked together in this formulation of the economy.

The OECD (2006) acknowledges the significant input of the U.K. government and U.K.-based think-tank Demos to its work on personalization. Personalization "springs from the awareness that 'one-size fits-all' approaches to school knowledge and organization are ill-adapted both to individual's needs and to the knowledge society at large" (p. 9). Through its focus on public sector reform, personalization promises to link "innovation in the public sector to the broader transformations in OECD societies" (p. 115). Personalization also challenges the teacher-learner relationship, placing the learner at the center. The teacher is now one amongst an army of specialists; a node in the network and drawn upon when necessary. The OECD report invites a new way of thinking about the learner when it asks:

> Imagine a catalogue that consists of items you invent, design and conceive yourself and the supplier was more of an assistant who connects up with you momentarily through a vast, continuously reconfigured network...In this post-industrial catalogue, which the "producer-consumer" or *prosumer* can publish as their personalised version others might want to build on, the crucial ingredient is the value added by the individual themselves. Their capacity to invent, design and then co-produce is what distinguishes this version of personalisation from mass customization. (p. 118)

In the United Kingdom, journalist Charles Leadbeater's writing on personalization has been extremely influential. In a pamphlet given government endorsement Leadbeater (2004) argues that it is possible to imagine that

> users take on some of the role of producers in the actual design and reshaping of the education system...The script of a system characterised by personal learning is rather different. It should start from the premise that the learner should be actively, continually engaged in

> setting their own targets, devising their own learning plans and goals, choosing from a range of different ways to learn. (P. 12)

This means breaking open education as the sole system of formal, institutionalized learning and moving toward one that is more fluid, flexible, multi-aged, and community-based (p. 16), and where teachers have a minor rather than major role.

Demos, whilst supportive of personalization as a project, is mindful of the challenges to its possible success. The costs of education are likely to be significantly greater than current mass systems, unless of course there are mechanisms for ensuring fairness whilst targeting the brightest talent. Even so, personalization will exacerbate the huge chasm between social classes as a result of differential access to cultural, economic, and social resources (Johnston, 2004). Personalization is intended to deliver neoliberalism without us knowing through its appeal to being more democratic as a result of involving us in the decisions we are making about services. However, personalization's neglect of poverty and inequalities, as well as its flawed assumptions about learners, autonomy, and learning, will likely render it a highly contradictory strategy in the knowledge economy armory.

Personalization articulates with notions of choice, individual responsibility, and risk, and the continual renovation of the self (Robertson, 2005). It takes the marketization of education a further stage, placing it at the very heart of the pedagogical process (Hartley, 2007, p. 630). There is a convergence, then, around the importance of human capital and learning into adulthood as part of an adjustment to the new economy and to promote social inclusion, and to invest in the future (Jenson and Saint-Martin, 2006). Personalization is envisaged as having the potential to be a mechanism of governance, a means of constituting the active subject, and co-constituting the competitive knowledge-based economy. It also introduces consumerism to education beyond policies of choice (where consumers made decisions between products). The consumer, in this case the learner, constructs the system, becoming in this moment both consumer and producer—a fluid, self-organizing model resonating with Castell's (1996) network society, and Bell's post-industrial futures imaginaries. However, personalization's success as a pedagogy for the knowledge-based economy will ultimately lie with whether it is capable of resolving multiple problems within the system of knowledge production—that is, if it is able to increase individual learner performance to ensure international competitiveness; generate sufficient self-discipline in the learner/worker; facilitate inclusion so that it is a bridge to self-responsibility; and generate creative minds to feed the innovations necessary for an economy centered on value from intellectual property.

The Biologization/Neurologization of the Learner

Brains feature a great deal in the various projects to realize a knowledge-based economy, from strategies to secure the best brains/talent from around the world to work for a firm or nation, to those that focus attention on how to "read" the brain so as to develop instructional approaches that nurture learning and creativity. Considerable attention is now being given to research on brains—though from the perspective of neuroscience. Its claim is that this kind of approach provides a "hard, scientifically based theoretical framework for educational practices . . . and the basis for a 'Science of Learning' " (OECD, 2007, p. 24).

Since 1999 the OECD's Centre for Education has run a program of work on the brain and learning in order to better understand the learning of an individual. The program has been developed over two phases. In phase one (1999–2002), an international group of researchers were bought together to review research findings on the brain and its implications for learning sciences. In phase two (2002–2006), three areas were further developed: literacy, numeracy, and lifelong learning. In its 2007 publication—*Understanding the Brain: The Birth of a Learning Science*, the OECD claims that through techniques such as "neuroimaging" it is possible to see extensive structural change taking place in the brain. With this kind of data the report claims that, for instance:

> Understanding the underlying developmental pathways to mathematics from a brain perspective can help shape the design of teaching strategies. Different instructional methods lead to the creation of neural pathways that vary in effectiveness: drill learning, for instance, develops neural pathways that are less effective than those developed through strategy learning. (P. 16)

Understandings generated from this approach to learning, such as the idea of *plasticity* (i.e., that development is a constant and universal feature of cerebral activity), is used to legitimize the lifelong learning discourses that feature as sub-narratives in the knowledge-economy master narrative.

However, this area of work has been particularly controversial, in part because of the huge (and often inaccurate) claims that have been made for brain research—in being able to understand processes learning (Hall, 2005, p. 4) and the considerable distance (still) between brain development, neural functioning, and education practices. As Bruer (1997) noted: "Neuroscience has discovered a great deal about neurons and synapses, but not nearly enough to guide educational practice" (p. 15).

The Commoditization of Schooling

A fourth project being mobilized is the unbundling and selective capitalization of the schooling system. This has been underway for some time in the heartlands of selected OECD countries—particularly the United States, the United Kingdom, New Zealand, and Canada. However, until recently, capitalization centered on the non-core aspects of education services (Molnar, 2006). Over the past five years it is possible to observe an extension and escalation of these activities, contributing in turn to a maturing and expanding education industry (Ball, 2007). Paralleling, though not directly propelling this development is the World Trade Organization (WTO) and its ongoing negotiations—to progressively liberalize the services sectors and bring them into the global trading regime (Robertson, Bonal, and Dale, 2002). This project's narrative is that the governance regime of knowledge-based economies should have a limited number of market-unfriendly policies (Robertson, 2009). Not only should state monopolies of public services—like health and education—be dismantled, but it is argued that the private sector is uniquely capable of managing change and innovation (Hatcher, 2006, p. 599).

Recently, there has been rapid overall growth in the commercialization/privatization of schooling as a result of both explicit government policies shaping the development of the sector, and also growing confidence by firms that profits can be made in particular areas of education services. Education as a sector is being unbundled to reveal an array of educational goods and services open to trade to market actors. This includes goods and services in areas such as (i) delivery—such as provision; (ii) content—such as texts; (iii) infrastructure—such as hardware, buildings; and (iv) services—such as testing. Unbundling is taking place in a number of sectors of the education system: K-12, higher education, and the corporate sector. However, my concern here is with K-12. A number of studies have recently been published to reveal the extent of the capitalization of education (see Ball, 2007; Hentscke, 2007; Mahony, Hextall, and Menter, 2005). Taken together they reveal a myriad of complex interconnections between firms that draw education directly into the global economy.

Education is now regarded as big business. Hentscke (2007) reports that in the United States for-profit firms operating in the K-12 segment had an annual growth rate of 6.6 percent (p. 178). The highest growth areas in the United States are currently in K-12 testing and tutoring, while growth in K-12 delivery has been propelled by the continuing expansion of Charter Schools, commercial home-school services, and virtual charter schools (p. 184). Expansion in the field of testing services also owes a great deal to the testing mandate imposed by the Bush administration—as a result of the effort to drive up standards in education to foster a more competitive U.S. economy.

The market is dominated by a small number of very large firms, such as Educate (previously Sylvan Learning Systems), Huntington Learning Centers, and Score! (Kaplan). Recently some of these firms in the United States have begun to buy up smaller firms both within and across national boundaries in order to generate economies of scale (Hentscke, 2007). New firms, such as Bairds, have also been created to provide advice to investors in the education sector. These firms monitor national and international education policy developments with an eye to which of these policies, such as with assessment policies, will provide an opening and opportunity for financial gain.

Similarly, in the United Kingdom the Labour government has looked to the private sector for the ongoing development of education. This was facilitated by the Private Finance Initiative (PFI)—legislation that enabled the private sector to move into hitherto uncharted territory for profit-making in education. These public-private initiatives have ranged from relatively small sponsorship deals to specialist schools and multimillion dollar infrastructure developments (such as building schools, taking over and managing local education authorities, school inspection, and examination marking). By 2004 there were PFI projects estimated by Treasury to be £7.7 billion, including £900 million for education projects (Ball, 2007, p. 46). Companies involved in the education service industry have also used a range of creative financing schemes to generate windfall profits—from renegotiating contracts to selling on the business to a new contractor (p. 167). Not surprisingly, the early days of the PFI/PPP was met with considerable resistance by teacher unions. However teachers lost considerable power over this period, as they moved from collective to individualized performance-based contracts. The outsourcing of education services in the United Kingdom has been controversial and in many cases inefficient—with companies facing annual fines for failing to meet targets, criticisms from the auditor general that quality is poor, and companies bailing out within a year or two of starting what were presumed to be fixed contrasts. The full public sector costs are therefore very difficult to estimate, but it is the public that bears the costs. Despite this, the government has pressed ahead and it is clear that this is an ideological project over and above all else. Ball (2007) describes in considerable detail the deepening interconnections between the state, capital, the public sector and civil society, arguing that the extent and consequences of these changes are epistemic; that is, they involve the reshaping of "deep social relations . . . in an emerging Market Society within which everything is viewed in terms of quantities; everything is simply a sum of value realised or hoped for" (p. 185).

Teachers' Matter—Don't They? Drawing Conclusions

The knowledge-based economy master narrative is a powerful one in its capacity to articulate with, and give direction to, projects, strategies, practices, and subjectivities that might underpin and realize a new long wave of accumulation. It ties education more closely and completely to the economy through prioritizing knowledge. However the price of this tie is that a more fundamental transformation of the education sector is required. This new order—a knowledge-based economy—requires and constitutes an ontological and epistemic shift in society. It embraces a very different way of thinking about what it means to be human, how humans develop, learn, and come to know the world. It also registers a different role for teachers, for good and for bad.

The current system of education, with its grammar created out of, and reflecting, education's role in the production of modernity and capitalism, is problematized in the various translation projects for the knowledge economy, as having now reached its "sell-by-date." The teacher as the secular bible must give ground to the learner and a new pedagogy of production. One reading the unfolding projects outlined earlier, of "modernization," "personalization," "scientization," "biologization/neurologization," and "commoditization," is that they assume a very different role for the teacher, because the learner is involved in a very different set of social relations The learner now replaces the teacher. However the pedagogical project for the learner is the making and remaking of goods and services for the economy in a continual process of re/invention and consumption.

The contradictions and dilemmas in these translations/strategies are all too evident. How can these approaches be sufficiently embedded for them to stabilize the social formation and economy? How is the necessary social cohesion to be built to ensure social stability and social reproduction? Taking teachers out of the formulation might remove an important obstacle to realizing the knowledge-based economy. However, keeping teachers in might be just as crucial. Teachers, if nimble and visionary, might also be well placed to realize their individual and collective interests when the inevitable contradictions generate new spaces for action.

References

Ball, S. (2002). *Class Strategies and the Education Market.* London: Routledge/Falmer.

———. (2007). *Education plc: Understanding Private Sector Participation in Public Sector Education.* London and New York: Routledge.

Bell, D. (1973). *The Coming of the Post-industrial Society: A Venture in Social Forecasting.* Middlesex: Penguin.

Blair, A. (2000). *Press Release—Knowledge Economy,* PM Speech, London.

Bruer, J. (1997). Education and the brain: A bridge too far. *Educational Researcher,* 26(8), 4–16.

Castells, M. (1996). *The Rise of the Network Society.* London: Blackwell.

———. (1999). Information Technology, Globalization and Social Development, UNRISD, Discussion Paper No. 114, September, Geneva, INRISD.

Cox, R. (1996). *Approaches to World Order.* Cambridge: Cambridge University Press.

Drucker, P. (1969). *The Age of Discontinuity: Guidelines to our Changing Society.* London: Heinemann.

Dale, R. (2008). Pedagogy and cultural convergence. In H. Daniels, J. Porter, and H. Lauder (eds.). *Routledge Companion to Education.* London and New York: Routledge. pp.

Dale, R. and Robertson, S. (2004). Interview with Boaventura de Sousa Santos. *Globalisation, Societies and Education,* 2(2), 141–146.

European Council. (2000). Conclusions of the European Lisbon Council, March 23–24, SN100/00; http: //www.europarl.europea.eu/summits/lis1_en.htm.

European Commission. (2007a). *Schools for the 21st century* (Commission Staff Working Paper). SEC (2007)1009. Brussels: European Commission.

———. (2007b). *Progress towards the Lisbon Objectives in Education and Training: Indicators and Benchmarks.* (Commission StaffWorking Paper). SEC (2007)1284, Brussels: European Commission.

Farnsworth, K. (2006). Business in education: a reassessment of the contribution of outsourcing to LEA performance. *Journal of Education Policy*, 21(5), 485–496.

Ferguson, I. (2007). Increasing user choice or privatising risk? The antinomies of personalization. *British Journal of Social Work*, 37, 387–340.

Florida, R. (2002). *The Rise of the Creative Class.* New York: Basic Books.

———. (2005). *Cities and the Creative Class.*.London and New York: Routledge.

Friedman, M. (1962). *Capitalism and Freedom.* Chicago: University of Chicago Press.

Hall, J. (2005). Neuroscience and education: what can brain science contribute to teaching and learning. *Spotlight*, 92, University of Glasgow.

Hargreaves, D. (2001). Teachers' Work. In OECD (ed.). *Knowledge Management for Learning Societies.* Paris: OECD.

Harvey, D. (2005). *A Brief history of neoliberalism.* Oxford: Oxford University Press.

Hatcher, R. (2006). Privatization and sponsorship: the re-agenting of the school system in England. *Journal of Education Policy*, 21(5), 599–619.

Hentscke, G. (2007). Characteristics of growth in the education industry: Illustrations from US education business. In K. Martens, A. Rusconi, and K. Leuze (eds.). *New Arenas of Global Governance: The Impact of International Organizations and Markets on Educational Policymaking.* New York: Palgrave, pp. 176–194.

Hartley, D. (2007). Personalisation: the emerging "revised" code of education? *Oxford Review of Education*, 33(5), 629–642.

ILO. (2004). A fair globalization: Creating opportunities for all. (The World Commission on the Social Dimensions of Globalization). Geneva: ILO.

Jenson, J. and Saint-Martin, D. (2006). Building blocs for a new social architecture: the LEGO™ paradigm for an active society. *Policy and Politics,* 34(3), 439–451.

Jessop, B. (2004). Critical semiotic analysis and cultural political economy. *Critical Discourse Analysis*, 1(1), 159–174.

Johnston, M. (2004). *Personalised Learning: An Emperor's Outfit.* London: DEMOS.

Leadbeater, C. (2004). *Learning about Personalization: How Can We Put the Learner at the Heart of the Education System.* London: DfES.

Lundvall, B-Å. (1992). *National Innovation Systems: Toward a Theory of Interactive Learning.* London: Pinter Publishers.

Machlup, F. (1962). *The Production and Distribution of Knowledge.* Princeton: Princeton University Press.

Mahony, P., Hextall, I., and Mentor, I. (2004). Building dams in Jordan, assessing teachers in England: a case study of edu-business. *Globalisation, Societies and Education,* 2(2), 277–296.

Mattelart, A. (2003). *The Information Society: An Introduction.* London: Sage

Molnar, A. (2006). The commercial transformation of public education. *Journal of Education Policy*, 21(5), 621–640.
OECD. (1996). *The Knowledge-Based Economy*. Paris: OECD.
———. (2000). *Schooling for Tomorrow Toolbox*. Paris: OECD.
———. (2001). *Knowledge Management for the Learning Society*. Paris: OECD.
———. (2005). *Teachers Matter: Attracting, Developing and Retaining Effective Teachers*. Paris: OECD.
———. (2006). *Personalising Education*. Paris: OECD.
———. (2007). *Understanding the Brain: The Birth of a Learning Science*. Paris: OECD.
Robertson, S. (2000). *A Class Act: Changing Teachers' Work, the State and Globalization*. London: Falmer/Routledge.
———. (2005). Re-imagining and rescripting the future of education: global knowledge economy discourses and the challenge to education systems. *Comparative Education*, 41(2), 151–170.
———. (2009). Producing knowledge economies: the World Bank, the KAM, education and development. In M. Simons, M. Olssen, and M. Peters (eds.). *Re-reading Education Policies: Studying the Policy Agendas of the 21st Century*. Dortrecht: Sense Publishers, pp. 251–274.
Robertson, S. and Dale, R. (2009). The World Bank, IMF and possibilities of critical education. In M. Apple (ed.). *Handbook of Critical Education*. London and New York: Routledge, pp. 23–35.
Robertson, S. and Lauder, H. (2001). Restructuring the education/social class relation. In R. Phillips and J. Furlong (eds.). *Education, Reform and the State*. Routledge/Falmer: London, pp. 222–236.
Robertson, S., Bonal, X., and Dale, R. (2002). GATS and the education service industry: The politics of scale and global territorialisation. *Comparative Education Review*, 46(4), 472–496.
Robertson, S., Novelli, M., Dale, R., Tikly, L., Dachi, H., and Alphonce, N. (2007). *Globalisation, Education and Development*. London: DfID.
Romer, P. (2007). Economic Growth. In D. Henderson (ed.). *The Concise Encyclopedia of Economics*. Liberty Fund., pp. 1–6.
World Bank. (2003). *Lifelong Learning for the Global Knowledge Economy*. Washington, DC: The World Bank.

Afterword

After Neoliberalism? Which Way Capitalism?

David Hursh

As I write this in early October 2008, the United States and much of the rest of the world face a financial crisis caused by several decades of neoliberal policies. While neoliberal proponents, such as Thomas Friedman (1999), have hijacked globalization to promote neoliberal policies by stating that, "the driving force behind globalization is free market capitalism" (p. 9), and therefore we have no alternative other than to promote deregulation, privatization, and markets, recent economic events reveal neoliberalism as a flawed and failed policy.

As minimally regulated investment banks greedily sought profits by making unwise loans that contributed to a housing bubble that finally burst, the finance industry faces economic collapse without the injection of federal funds. In response, the Bush administration and Treasury secretary Henry Paulson decided that government is not, after all, the problem but, instead, part of the solution, and proposed to use seven hundred billion dollars in public funds to bail out the country's financial institutions. However, Paulson's proposal was defeated as legislators balked at a plan that granted Paulson absolute authority to do whatever he wanted.

Instead, some, including presidential candidate Barack Obama, have pointed out that the fiscal crisis results from the lack of governmental regulations, and that we need to rethink the relationship between corporations and government. Others, myself included, would like to use the crisis to replace neoliberalism with a form of social-democratic liberalism. Naomi Klein (2008), for example, writing in *The Guardian,* asks if the government can afford eighty-five billion dollars for the insurance giant AIG. She asks, "Why is single-payer health care—which would protect Americans from the predatory practices of heath-care insurance companies—seemingly such an unattainable dream?" (September 19).

Katrina Vanden Heuvel and Eric Schlosser, the former the publisher of *The Nation,* in an article in the *Wall Street Journal* (September 27, 2008), specifically call for "a new New Deal: a systemic approach to the financial and economic problems of the United States," that provides "relief, reform, and reconstruction." They endorse Hillary Clinton's call for a revival of the Home Owners Loan Corporation (HOLC), which in the New Deal reduced the number of "foreclosures by purchasing troubled mortgages from banks and then reissuing them with more favorable terms." They call for regulatory reform so that we are protected from unethical and greedy actions of corporations. Last, they call not only for reconstruction of the nation's physical structure but also for providing universal health care and investment in renewable energy and public-works projects.

But neoliberalism will not die quietly and neoliberals will see the crisis as a chance to push harder for markets and privatization. Klein (2008, September 19) writes that:

> The massive debts the public is accumulating to bail out the speculators will then become part of a global budget crisis that will be the rationalization for deep cuts to social programs, and for a renewed push to privatize what's left of the public sector. We will be told that our hopes for a green future are, sadly, too costly.

Whether global neoliberalism can be pushed back and what a new social-democratic liberalism emphasizing a socially just and environmentally sustainable future would look like remains to be seen. The contributors to this volume help us understand the difficulties we face. Jill Pinkney Pastrana describes Chile's long road back toward democracy decades after Dictator Augusto Pinochet's neoliberal policies that weakened and almost destroyed the public sector. Salim Vally, Enver Motala, and Brian Ramadiro bravely take on the neoliberal policies of the postapartheid South African government. Rather than undertaking the reforms necessary to create a more democratic and equitable society, the African National Congress (ANC), under pressure from global financial institutions, instituted policies that benefited the old and new elite rather than the working class. After visiting the Republic of South Africa in 1999, I wrote that "while democracy has come to South Africa, equality has not" (Marais, 1998, p. 245). Vally, Motala, and Ramadiro show us how community-oriented participatory action research can work to reveal and counteract the devastating effects of neoliberal policy.

Furthermore, the contributors to this volume show how education and economic policies are always intertwined. Leistyna shows us how education policies promote corporate growth either through the privatization of schools or through the promotion of for-profit curricular materials, and Smyth suggests how we might incorporate critical pedagogy to critique neoliberalism.

The question remains unanswered: Which way capitalism? Do neoliberals use the current economic crisis to promote an even more pernicious form of neoliberalism in which public institutions and property, encompassing everything from schools to roads, are sold off to the highest bidder? Or do we reinstate a form of social democratic liberalism in which the public welfare and the common good take precedence over private profit? The fiscal crisis poses both great possibilities and dangers.

I close my most recent book, *High-stakes Testing and the Decline of Teaching and Learning* (2008), by calling for a new social movement in which educators work with others to combat high-stakes testing, the rise of markets, and privatization in education, and indeed, in all of our social policies, as well as the neoliberal thinking that supports those policies. We need to create a society in which everyone has access to health care, a useful and decent paying job, adequate housing, and a good education. Moreover, we need to create a society in which achieving these goals is central to our political and economic policies and not left to the vagaries of the market, which we now realize is no assurance at all. We need, as the authors point out, to create a new society based on principles of environmental and social justice.

References

Friedman, T. (1999). *The Lexus and the Olive Tree.* New York: Farrar, Straus & Giroux.

Hursh, D. (2008). *High-states Testing and the Decline of Teaching and Learning: The Real Crisis in Education.* New York: Roman and Littlefield.

Klein, N. (2008, September 19). Free market ideology is far from finished. *The Guardian.* Retrieved September 22, 2008, from http:www.guardian.co.uk/commentisfree/2008/sep/19/marketturmoil.usa.

Marais, H. (1998). *South Africa Limits to Change: The Political Economic of Transformation.* Cape Town: University of Cape Town Press.

Vander Heuval, K. and Schlosser. (2008). America Needs a New New Deal. *Wall Street Journal, September 27, 2008,* http://www.pnhp.org/news/2008/september/america_needs_a_new_.php.

Contributors

Wayne Au is an assistant professor of secondary education at California State University, Fullerton. He most recently published *Unequal by Design: High-Stakes Testing and the Standardization of Inequality* (Routledge, 2008). He is also an editor for the journal *Rethinking Schools*, and his research interests include social justice education and critical educational theory.

Dave Hill is professor of education policy at the University of Northampton, England, and professor of education at Middlesex University, London, England. He is a longtime socialist labor union and political activist and former elected representative/local/regional leader. He lectures worldwide on class, race, education policy and resistance. He is founder editor of *The Journal for Critical Education Policy Studies*, www.jceps.com.

David Hursh, associate professor in teaching and curriculum at the Warner School of Education at the University of Rochester. His recent publications and political efforts focus on the relationship between neoliberal economic and education policies. Since 1999, he has been on the steering committee of the Coalition for Common Sense in Education, a group working to repeal high-stakes testing. His most recent book is *High Stakes Testing and the Decline of Teaching and Learning: The Real Crisis in Education* (Rowman and Littlefield, 2008).

Sebastjan Leban is an artist and theoretician from Ljubljana. His artistic practice involves the collaboration with Staš Kleindienst, the group Trie, and the group Reartikulacija. He is one of the founders and the editor of the journal *Reartikulacija* and has exhibited in numerous national and international exhibitions, participated in many symposiums and lectures, and published texts in several different publications.

Pepi Leistyna is an associate professor in the Applied Linguistics Graduate Studies Program at the University of Massachusetts, Boston. He coordinates the research program and teaches courses in cultural studies, media analysis, critical pedagogy, and language acquisition. Speaking internationally on issues of democracy, education, and social justice, Leistyna's books include *Breaking Free: The Transformative Power of Critical Pedagogy* (Harvard Publishing Group), *Presence of Mind: Education and the Politics of Deception* (Westview Press), *Defining and Designing Multiculturalism: One School System's Efforts* (SUNY Press), *Cultural Studies: From Theory to Action* (Blackwell Press), and *Corpus Analysis: Language Structure and Language Use* (Rodopi Press). His forthcoming book is entitled *Laughing Matters: Entertainment Television's Mockery of the Working Class.*

Sheila Macrine, PhD, is literacy professor in the Deborah Cannon Partridge Wolfe College of Education, New Jersey State University, New Jersey. Her research focuses on connecting cultural, political, and feminist frameworks to institutional and personal contexts of pedagogy, particularly as they relate to the social imagination, progressive democratic education, and critical studies. She has published recent scholarly articles and chapters on inclusion, neoliberalism in education, the impact of No Child Left Behind, and instruction and assessment in the literacy classroom.

Peter McLaren, PhD, is a professor at the Graduate School of Education and Information Studies, University of California, Los Angeles. He is the author, coauthor, editor, and coeditor of approximately forty books and monographs. Several hundred of his articles, chapters, interviews, reviews, commentaries, and columns have appeared in dozens of scholarly journals and professional magazines since the publication of his first book, *Cries from the Corridor*, in 1980 (Methuen, Toronto; New York). His work has been translated into seventeen languages. He lectures internationally and is a member of the Industrial Workers of the World.

Martha Montero-Sieburth is currently a research fellow at the Institute for Migration and Ethnic Studies at the University of Amsterdam, where she is completing several manuscripts on Latin Americans in Spain and Latinos in the United States. She is a professor emerita of the Department of Leadership in Education at the Graduate College of Education at the University of Massachusetts-Boston where she taught in the Leadership in Urban Schools Doctoral Program and the Educational Administration Masters Program until August 2007. Her ethnographic research has been conducted in U.S.

urban schools and communities, and in greater metropolitan areas of Boston where she has studied the policies and practices toward Latino adolescents in secondary schools, Latino parental involvement, community ethnography of Latinos, and the leadership of Mexicans in the New England area.

Enver Motala was a lawyer for the independent trade union movement and also played a significant role in the antiapartheid education movement. After the first democratic elections he was appointed the deputy director-general of education in the province of Gauteng. He is presently an associate of the Education Policy Consortium for whom he has coordinated research projects on democracy, human rights, and social justice in education in South Africa. He has also done similar work for the Nelson Mandela Foundation.

João M. Paraskeva, PhD, is a professor of educational leadership at University of Massachusetts, Dartmouth, MA. Prior to his appointment at Umass, Dartmouth, he was a visiting professor of educational leadership at University of Miami, Ohio. He also taught for ten years at the University of Minho in Portugal. He is an international scholar and has published twelve books, as well as numerous chapters and articles, on a wide range of topics in the field of critical education. As a result, Paraskeva has established himself as an important critical scholar.

Jill Pinkney Pastrana is an associate professor of Foundations of Education at the University of Wisconsin-Eau Claire. She was recently named a 2008–2009 Fulbright Scholar, and this will allow Pastrana to work with colleagues at the Catholic University in Valparaiso in Chile. Pastrana's areas of expertise include international and comparative education, multicultural and intercultural education, language minority education, politics and education, and globalization and education policy reform.

Brian Ramadiro is the deputy director of the Nelson Mandela Institute for Education and Rural Development at the University of Fort Hare. He is the former co-coordinator of the ERP. His scholarship focuses on youth issues, "race," culture, and indigenous knowledge. His current projects include the promotion of bilingualism and multilingualism in education, rural development, and methodologies for community mobilization and participation in public policy planning and development.

Susan L. Robertson is a professor of sociology of education in the Graduate School of Education, University of Bristol. Along with her colleague Roger Dale, she is the founding editor for the journal *Globalisation, Societies and Education* published by Carfax. Robertson recently completed a Synthetic Review of Globalization, Education and Development for the Department of International Development. Her current work is engaged with globalization and regionalization as it works on and through both education systems and new sites of knowledge production.

John Smyth is a research professor of education at the University of Ballarat. He is author/editor of fifteen books including, most recently, *Teachers in the Middle: Reclaiming the Wasteland of the Adolescent Years of Schooling* (with Peter McInerney, Peter Lang Publishing, 2007) and *"Dropping Out" Drifting Off, Being Excluded: Becoming Somebody Without School* (with Robert Hattam and others, Peter Lang Publishing, 2004). His research interests include policy ethnographies of schooling, issues of social justice, community renewal, and policy sociology of students' lives and teachers' work.

Salim Vally worked as a teacher and trade unionist before joining the Wits Education Policy Unit. He is active in social justice and solidarity movements and is a board member of various professional organizations. He is involved in scholarship that combines interdisciplinary and comparative approaches to critically examining education policy and practice. His abiding interest is in encouraging his peers and community activists to collaborate on socially engaged research. Salim was visiting scholar at York University, Ontario, from 2007 to 2008 and has resumed his post as senior researcher at the Wits EPU and the coordinator of the ERP.

Richard Van Heertum is currently a visiting assistant professor in the education departments at CUNY/College of Staten Island teaching classes in foundations and the interdisciplinary Core program. He recently completed his PhD in education and cultural studies at the University of California, Los Angeles, where his dissertation focused on cynicism and democracy. He also has an MA in economics. His work has appeared in *Policy Futures in Education, Interactions, McGill Journal of Education*, and a number of anthologies, and he has published extensively in the popular press. He previously served as the program officer for the Paulo Freire Institute at UCLA.

Index